THE COMMUNITY BANK AI SERIES

The Community Bank AI Governance Handbook

VOLUME I

The Definitive Guide to AI Risk, Oversight & Compliance

JOHN HUMPHREY JR.

For Dad

John Humphrey

1964 – 2021

Who taught me to put faith and family before all else.

I hope I make you proud.

The Community Bank AI Governance Handbook

VOLUME I

The Definitive Guide to AI Risk, Oversight & Compliance

✦

LEGAL DISCLAIMER

This book is intended for informational and educational purposes only. It does not constitute legal, regulatory, financial, or professional advice. The information, materials, and recommendations provided herein are based on the author's professional experience and interpretation of publicly available regulatory guidance as of the date of publication. They should not be relied upon as a substitute for consultation with qualified legal counsel, regulatory advisors, compliance professionals, or other subject matter experts.

Reading this book or using its materials does not create an attorney-client, advisory, consulting, or other professional relationship with the author or Evenstar Advisory Group LLC.

The regulatory landscape governing artificial intelligence in banking is evolving rapidly. Laws, regulations, supervisory guidance, and examination practices referenced in this book may have changed since publication. Readers are responsible for verifying the current status of all regulatory requirements applicable to their institution and for obtaining professional advice tailored to their specific circumstances.

The policy templates, assessment tools, questionnaires, and other materials in the appendices are designed for practical adoption. They should be reviewed by qualified professionals familiar with the adopting institution's specific regulatory environment, risk profile, and organizational structure. The appendices present completed worked examples; the AI Governance Toolkit provides blank, editable versions for implementation.

No representation is made that use of these materials will result in any particular examination rating, supervisory outcome, or regulatory result.

This book references or links to third-party websites, publications, products, and services, including government and regulatory agency websites, that are not under the control of the author or Evenstar Advisory Group LLC. Such references are provided for convenience and information only and do not constitute endorsement or recommendation. The author and Evenstar Advisory Group LLC are not responsible for the content, accuracy, availability, or continued operation of any external resource. Web addresses cited were current as of the date of publication and may change or cease to function thereafter.

✦

TRADEMARK ACKNOWLEDGMENTS

The Community Bank AI Series, Volume I
First Edition, 2026

Published by Evenstar Advisory Group LLC

ISBN 979-8-9956437-1-5 (hardcover)
ISBN 979-8-9956437-0-8 (paperback)

Edited by Jessica Filippi

Cover and interior design by Silke Spingies

Acknowledgments

A book like this accumulates debts. These are mine, and I am glad to name them.

To my loving mother, Kimberly, and my wonderful little sisters, Christina and Maria—this book wouldn't exist without you three. Thank you for all those days and nights of listening to draft after draft, and for helping me find the words. This book is ours.

To Terri Brinkman, who trusted my judgment and gave me room to grow. I'm grateful for that confidence.

To Riley Adams, for honoring this book with a foreword that reflects both the depth of your experience and the clarity of your perspective. Your voice carries weight in this industry, and I am grateful you lent it here.

To Jessica Filippi, who edited these pages with a sharp eye and a light hand, and to Silke Spingies, whose design gave them their shape—thank you both for the craft you brought to this book.

And to the community bankers who will read this book and put it to work—this was written for you. The trust your communities place in you is something worth protecting, and I hope these pages help you do that.

Foreword

"A defining first resource of its kind—your compass for direction and your legend for clarity in AI deployment."

In the world of finance, we often speak about globally systemically important banks as if they operate in a different universe. Having spent my career as both a regulator and an executive in those environments, I understand the scale and complexity they manage. But I have also learned a simpler truth: regardless of size, every bank is built on the same foundation—trust, discipline, and a commitment to serve its community. Artificial intelligence is now testing that foundation.

What makes this moment different is not simply the speed of technological change—it is the nature of the decision in front of every community bank. The industry is facing a two-sided risk. On one side is the cost of inaction: falling behind in efficiency, relevance, and competitiveness. On the other is the cost of unmanaged adoption: regulatory exposure, unintended bias, data risk, and erosion of customer trust. The path forward is not to choose between these risks but to govern them. Done well, governance ensures decisions remain on point, balanced, and aligned with institutional values. That is the focus of this book.

John Humphrey understands something that many conversations about AI miss: the challenge is not whether community banks will use artificial intelligence—they already are. The real question is whether they will govern it effectively before it begins to shape decisions beyond their control. This handbook does not treat governance as a compliance exercise or a technical afterthought. It positions governance as the mechanism that enables confident adoption, sustainable growth, and responsible innovation.

At the institutional level, we have long understood that data discipline underpins sound risk management. That principle applies just as powerfully here. But this book goes further. It connects data integrity to accountability, to fairness in outcomes, to explainability for customers and regulators, and ultimately to the preservation of trust. These are not abstract concepts—they are operational expectations that must be built into how AI systems are selected, implemented, and overseen. Importantly, this is an area where examiners and regulators can appreciate and embrace a tactful, judicious approach—one that balances innovation with control and progress with prudence.

What makes this work especially valuable is its practical orientation. John does not leave the reader at the level of principle. He provides clear, actionable frameworks—how to inventory AI systems, how to classify risk, how to structure governance oversight, how to manage vendors, and how to report to boards and regulators. For community banks, this is essential. Governance must be proportional, practical, and executable with the resources available.

The book also recognizes something that deserves emphasis: community banks are not simply smaller versions of larger institutions. They are fundamentally different. They operate on relationships, local knowledge, and direct accountability to the communities they serve. The decisions informed by AI are not abstract—they affect people you know, businesses you see every day, and communities you are part of. That proximity is not a disadvantage; it is a strength. It is what makes thoughtful governance not just a regulatory necessity but a reflection of the values that define community banking.

As we look ahead, we will continue to see advancements in automation and increasingly autonomous systems. But no matter how sophisticated the technology becomes, the responsibility remains human. Governance defines where that responsibility lives, how it is executed, and how institutions ensure that technology serves—not replaces—the judgment and integrity at the core of banking.

If there is one takeaway from this work, it is this: institutions that succeed with AI will not be those that adopt it fastest, but those that adopt it with clarity, discipline, and purpose. This book provides the roadmap to do exactly that.

Riley Adams
Chief Risk Officer - First Resource Bank

About This Book

Artificial intelligence reached community banks ahead of most of their governance programs, and it is still accelerating. Machine learning models already score loans and flag fraud. Generative AI drafts customer communications and credit memos. Agentic systems that act on their own are arriving on vendor release schedules, with or without an adoption decision. Most governance guidance, meanwhile, is written for the largest institutions—leaving community banks to shrink enterprise frameworks they were never meant to staff.

This book closes that gap. It is a complete, practical governance program built for a community bank's scale and for the three groups who will actually run it: the IT leaders and CISOs who implement and secure AI systems, the governance and risk committees who oversee and challenge them, and the boards of directors who answer for the risk.

Each chapter pairs strategic principles with working tools—policy templates, assessment checklists, meeting agendas, reporting structures, and decision matrices—every one of them completed in the appendices and provided as editable files in the companion Community Bank AI Governance Toolkit. And every page applies one test: proportionality. The governance in this book fits the size, complexity, and risk profile of the institution you actually lead.

WHO IS THIS BOOK FOR?

This book is for the people who carry responsibility for AI decisions at community banks—whether they chose that responsibility or it landed on their desk one Monday morning. If any of the following describes you, this book is yours.

Board Members and Directors who know AI is on the agenda but aren't sure what questions to ask, what reports to demand, or where fiduciary responsibility begins and ends. Your job is the risk, the oversight, and the questions that matter; the algorithms have their own specialists. Chapters 13 and 17, the board reference card in Appendix H, and the case studies will become your most-used pages.

Governance and Risk Committee Members who are tasked with overseeing AI adoption and need more than a mandate: meeting agendas, decision criteria, dashboards, and challenge questions. Chapters 6, 8, and 12 are your operating manual. The completed appendix examples show you the standard to aim for.

IT Leaders, CISOs, and Technology Officers who are building, managing, and securing AI systems while explaining them to non-technical colleagues. You live at the intersection of implementation and governance, and you need both the controls and a way to translate them upward. Chapter 10 supplies the controls; Chapters 7, 9, and 11 supply the governance tools for the policies, the vendors, and the agents.

CEOs and Senior Executives who set direction without getting lost in technical detail. You need to know what to fund, what to prioritize, what to ask your IT leader, and what to tell your board. The Quick Start gets you moving in 90 days; the strategy tools in Chapters 13 and 17 tell you when to invest and when to wait.

Compliance Officers and Risk Managers who watch AI risk land in their portfolio and need to know how the existing rulebook—SR 26-2 (formerly SR 11-7), fair lending, GLBA, third-party risk—applies to systems they didn't select and can't see inside. Chapter 4 maps the landscape; Chapter 5 turns bias testing and fair lending analysis into procedure.

Internal Auditors who must add AI to the audit plan—and anyone whose next examination will ask about it. Chapter 14 provides the methodology, including a scoped first-year audit, and the worked appendices show what auditable governance documentation looks like.

If you work at a community bank and AI has entered your world—whether through a vendor upgrade you didn't ask for, an employee's ChatGPT habit you just discovered, or a board member's question you couldn't fully answer—this book was written for you.

HOW TO USE THIS BOOK

Chapters carry audience tags (IT Leaders, Governance Committees, Boards) while remaining useful to every reader. The book runs in six parts, from foundations to what comes next. Read it straight through for the full education, or use the role table below to jump straight to the chapters that match your chair.

YOUR ROLE	START HERE	THEN READ
CEOs & Senior Executives	Quick Start, Ch. 1, Ch. 17	Ch. 6, Ch. 13, Case Studies
Board Members	Quick Start, Ch. 13, Appendix H	Ch. 2, Ch. 15, Ch. 17, Case Studies
Governance Committee Members	Quick Start, Ch. 6, Ch. 8	Ch. 12, Ch. 7, Appendix B
IT Leaders / CISOs	Quick Start, Ch. 10, Ch. 7	Ch. 8, Ch. 9, Ch. 11
Compliance Officers	Ch. 4, Ch. 5, Appendix I	Ch. 7, Ch. 9, Ch. 15
Internal Auditors	Ch. 14, Appendix B	Ch. 8, Ch. 7, Appendix F

From Reading to Implementation

The appendices show the program's core documents as completed worked examples, drawn from a fictional community bank to illustrate what good looks like in practice. The AI Governance Toolkit provides the program's documents as blank, editable files—Word documents, Excel workbooks, and PowerPoint decks your team can complete and deploy. See the Toolkit page at the back of this book for details.

Table of Contents

II THE RULES THAT ALREADY APPLY

IV MANAGING THE RISKS

VI PUTTING GOVERNANCE TO WORK

POLICIES, TOOLS & RESOURCES

"The beginning is the most important part of the work."

— PLATO, *THE REPUBLIC* —

✦ QUICK START ✦

Your First 90 Days of AI Governance

SIX ACTIONS THAT MOVE YOU FROM ZERO TO FOUNDATIONAL IN ONE QUARTER

Start building before you finish reading the handbook. The six actions in this Quick Start guide are the critical path—the moves that buy the most protection in the least time. Execute them in order over the next 90 days, and you will go from ungoverned AI to a Foundational governance program ready for its first examination conversation—with a documented path beyond it, protection for your customers, and grounds for your board to approve the next AI investment with confidence.

Everything else in the book builds on what you accomplish here. The AI Governance Toolkit includes a 90-Day Quick Start Tracker to plan and track these six actions, plus editable templates for every document the six actions produce.

WEEK	ACTION	WHAT TO DO	REFERENCE
1–2	Discover	Build your AI inventory. Survey vendors, employees, and technology platforms. Assess the data foundation for each system—source, quality, ownership, and gaps.	Ch. 8 (Sec. 8.1), Appendix F
3–4	Classify	Score each system across five risk dimensions. Assign risk tiers.	Ch. 8 (Sec. 8.2), Appendix B
5–6	Govern	Form the AI Governance Committee. Hold the first meeting.	Ch. 6 (Sec. 6.2), Appendix E
7–8	Policy	Adopt the AI Governance Policy and Generative AI Acceptable Use Policy. Get board approval.	Ch. 7 (Sec. 7.1, 7.7), Appendix A, G
9–10	Assess	Complete risk assessments for all High- and Critical-tier systems.	Ch. 8 (Sec. 8.3), Appendix B
11–12	Report	Deliver first quarterly AI governance report to the board.	Ch. 13 (Sec. 13.2), Appendix D

Action 1: Build Your AI Inventory (Weeks 1–2)

Why this comes first: You cannot govern what you do not know you have. First inventories always come back longer than expected. Vendor-embedded features, employee use of generative AI tools, and capabilities activated through routine software updates add up to a footprint larger than anyone in your organization realizes.

WHAT TO DO

1. Contact every technology vendor and ask directly: "Does your product use artificial intelligence, machine learning, or automated decision-making in any capacity?" Track responses in Appendix F's inventory format. Send the letters in week one and keep moving—vendors routinely take 30 to 90 days to respond, so build the inventory from what you can see and fold vendor confirmations in as they arrive.

2. Send a brief, non-punitive survey to all employees asking what AI tools they use for work, including personal tools used on bank devices or for bank purposes. Make clear that nobody is in trouble—you are counting, and amnesty gets honest answers. When the survey surfaces unmanaged generative AI use—and it will—do not wait for the formal policy: have management issue a one-page interim acceptable-use memo now, setting the ground rules for what can and cannot go into AI tools. The policies you adopt in Action 4 supersede it. The AI Governance Toolkit provides a ready-to-send Interim Acceptable-Use Memo template for exactly this step.

3. Review your technology environment—network traffic logs, software inventory, cloud subscriptions—for AI tools that were not reported.

4. Compile your AI inventory. See Appendix F for a completed example. You will be surprised by the volume. That is normal.

5. For each AI system you identify, assess the data foundation it depends on. Treat this step as load-bearing: dirty data produces dirty AI, and no governance program can fix outputs built on a fragmented data foundation. For every system in your inventory, document where the data comes from, how it flows into the AI system, who owns it, whether it is accurate and current, and whether it arrives in real time or through batch files that may already be stale by the time the model processes them.

Look for the warning signs: critical data trapped in legacy cores, disconnected systems, and inconsistent spreadsheets that nobody reconciles. If your AI systems are pulling from fragmented silos—isolated databases or systems that do not share data with each other—rather than unified, standardized data layers, your models will turn that fragmentation into errors at machine speed.

Treat data quality as a strategic finding. Flag every data gap you discover in the inventory and carry those findings into Action 5 and your first board report. Appendix K contains a Data Readiness Scorecard, and Chapter 10 (Section 10.2) explains how to interpret the results.

> **Expected Result**
> A documented inventory of every AI system in your organization, including vendor-embedded systems and employee-adopted tools. First inventories at community banks typically surface 15–40 systems. You are now ahead of most of your peers.

Action 2: Classify Every System by Risk (Weeks 3–4)

Why this matters: The inventory tells you what you have. Classification tells you where to focus. The four-tier system—Minimal, Moderate, High, Critical—concentrate your attention where the risk lives. Most systems will land at Minimal or Moderate, requiring only lightweight oversight. The few that land at High or Critical will demand your full attention.

WHAT TO DO

1. Score each system in your inventory across five dimensions using the risk scoring matrix in Chapter 8 (Section 8.2): Customer Impact, Decision Autonomy, Regulatory Sensitivity, Data Sensitivity, and Model Complexity.

2. Assign the overall risk tier based on the highest dimension score.

3. Flag every High and Critical system—these are your governance priorities.

Do not agonize over borderline cases. When in doubt, round up. You can reclassify later as your understanding deepens.

> **Expected Result**
> Every AI system in your inventory now has a risk tier. You can see, at a glance, where your most significant AI risks are concentrated. Typically, community banks find 2–5 High or Critical systems, 5–10 Moderate systems, and the remainder are Minimal. Your attention now has a short list.

Action 3: Form the AI Governance Committee (Weeks 5–6)

Why a committee first: governance is decisions, and decisions need a forum. The committee is the cross-functional body—IT, compliance, risk, and business lines—that reviews the policies you will adopt in Action 4, oversees the inventory you built in Actions 1 and 2, and owns the program from here forward. Policies without a body to enforce, review, and evolve them are paper exercises.

Forming the body first also means your policies reach the board with committee review behind them—the recommendation trail examiners look for.

No time or resources for a new committee? Many community banks have the IT Steering Committee absorb the AI Governance Committee's responsibilities instead—either permanently or as a starting point until a standalone committee is warranted.

If you take that route, amend the IT Steering Committee's charter to add the AI governance mandate—policy review and recommendation, inventory oversight, board reporting—and have the board approve the amendment. The responsibilities section of the Appendix E charter lifts directly into an existing charter. One check: confirm the membership includes compliance, risk, and a business-line voice alongside technology.

WHAT TO DO

1. Draft your AI Governance Committee Charter—or the charter amendment for the committee absorbing the mandate. See Appendix E for a completed example.

2. Identify five to seven members—three are enough to start: IT, compliance/risk, and a business line representative. Add a senior executive sponsor.

3. Get board approval for the charter or charter amendment.

4. Schedule the first meeting. Use the model agenda from Chapter 12 (Section 12.1). The first meeting should review the AI inventory, confirm risk classifications, set the schedule for reviewing the policies you will draft in Action 4, and identify the top governance priorities for the next quarter.

> **Expected Result**
> A chartered, board-authorized governance committee—standalone or absorbed into the IT Steering Committee—that has met at least once and established its operating rhythm. The governance program now has an engine.

Action 4: Adopt Your First Two Policies (Weeks 7–8)

Your bank needs a full policy suite—Chapter 7 walks you through all seven policies—but you do not need all seven to start governing. Two cover the most ground immediately. The AI Governance Policy establishes the scope, roles, risk tiers, and decision rights. The Generative AI Acceptable Use Policy sets the boundaries: shadow AI, data handling, and the ChatGPT window your loan officer has open right now. Shadow AI has not been waiting—the interim memo from Action 1 has been holding the line. These policies replace it with board-approved rules.

WHAT TO DO

1. Draft your AI Governance Policy and your Generative AI Acceptable Use Policy. See Appendices A and G for completed examples.
2. Present the drafts to the AI Governance Committee for review and approval.
3. Present to the board for approval with the committee's recommendation. The board presentation should take fifteen minutes: here is what AI we are using, here are the risks, here is the policy to manage them, here is our plan to build a full governance program.
4. Communicate the policies to all employees with a brief training session or memo explaining the key rules—particularly what data can and cannot go into AI tools.

> **Expected Result**
> Board-approved policies, recommended by your governance committee, that establish clear boundaries for AI use across the organization. Shadow AI now has an answer: here are the rules. Employees have clarity. Examiners have evidence of governance intent.

Action 5: Assess Your Highest-Risk AI Systems (Weeks 9–10)

Why now: Your inventory and classification identified the systems that carry the most risk. Before your first board report, you need documented risk assessments for these systems that demonstrate you understand the risks and have a plan to manage them.

WHAT TO DO

1. For each High- and Critical-tier system, conduct an AI risk assessment. See Appendix B for a completed example.
2. Involve the system owner, IT, and compliance in each assessment. For vendor systems—most of them, at a community bank—request what only the vendor can supply: model documentation, training data answers, and bias and performance testing results. Chapter 9 and Appendix C give you the question set.
3. Identify the most significant risks and the controls currently in place.
4. Document a monitoring plan for each system: What KPIs will you track, how often, and what triggers escalation?
5. Present completed assessments to the governance committee for review and approval.

> **Expected Result**
> Documented, committee-approved risk assessments for your most significant AI systems. You can now articulate, to an examiner or the board, what risks each system presents and how you are managing them. The systems that worried you most are now the systems you know best.

Action 6: Deliver Your First Board Report (Weeks 11–12)

Why this closes the loop: The board bears ultimate fiduciary responsibility for AI risk. Your first quarterly report establishes the reporting cadence, gives the board the information it needs, and creates the documented governance record that examiners expect.

WHAT TO DO

1. Prepare your first Board AI Report using the information gathered in Actions 1–5. See Appendix D for a completed example.
2. Present to the board: here is our AI inventory, here are the risk classifications, here is the committee we've formed, here are the policies we've adopted, here are the assessments of our highest-risk systems, and here is our plan for the next quarter.
3. Include a recommendation for next steps—typically completing the remaining policy suite (Chapter 7), expanding monitoring, and beginning vendor AI due diligence (Chapter 9).
4. Document the board's discussion and any questions or direction in the minutes.

> **Expected Result**
> A documented quarterly board report that establishes the reporting cadence and demonstrates governance engagement at the highest level. You have now achieved Foundational maturity on the roadmap in Chapter 17.

WHAT COMES NEXT

In 90 days, you have built an AI inventory, mapped the data behind each system, classified every system by risk, formed a governance committee, adopted foundational policies, assessed your highest-risk systems, and delivered your first board report. That is a defensible, documented governance foundation—still early, and already enough to demonstrate to examiners, to your board, and to your community that your bank takes AI governance seriously.

From here, the roadmap is clear. Over the next two quarters, complete the full policy suite (Chapter 7), build out vendor AI governance (Chapter 9), implement ongoing monitoring (Chapter 8, Section 8.4), and deliver training across the organization (Chapter 16). Within a year, you will have reached Evolving maturity. Within two years, Established. The chapters ahead give you everything you need to get there.

Open Appendix F, send the vendor inquiries, and build your inventory. The rest will follow.

PART I

Foundations

"Every community bank is already using artificial intelligence somewhere. The open question is whether you will govern it before it governs you."

— COMMUNITY BANKING EXECUTIVE,
MIDWEST REGIONAL CONFERENCE, 2025 —

✦ CHAPTER 1 ✦

The AI Moment for Community Banks

WHY GOVERNANCE CAN'T WAIT—AND WHY COMMUNITY BANKS HAVE UNIQUE ADVANTAGES

On a Tuesday morning in the spring of 2024, a community bank president in the upper Midwest received three pieces of news within an hour. First, his chief lending officer mentioned that a fintech competitor had just launched an AI-powered small business lending product that could approve loans in under ten minutes. Second, his compliance officer flagged a regulatory bulletin highlighting supervisory expectations for banks using artificial intelligence in consumer-facing products. Third, his IT manager forwarded a vendor email promoting an AI-driven fraud detection upgrade embedded in the bank's core processing platform—already switched on by default.

None of these developments were hypothetical. All three were happening in real time, and all three demanded governance decisions the bank was not yet equipped to make. This president's experience is now the norm. The only unusual thing about his morning was the compression: competition, supervision, and software that updates itself now arrive at every community bank—rarely in the same hour.

Artificial intelligence has arrived at community banks as a present-day reality—woven into the software, platforms, and services banks already use. The remaining question is how to engage it responsibly, safely, and in keeping with the values that define community banking.

This chapter establishes the landscape. We will explore why community banks occupy a unique position in the AI adoption curve, examine the real costs of both inaction and unmanaged adoption, survey the regulatory environment, and introduce the central argument of this book: governance is the foundation that makes sustainable AI innovation possible. Chapter 2 then defines what counts as AI in a community bank context—and shows where it is already at work in your bank.

1.1 The Community Bank Difference

Community banks occupy a distinct position in the financial system, and that position carries both unique challenges and unique advantages for AI governance. Building a program that actually works begins with those differences; a program that merely imitates one designed for institutions with fundamentally different characteristics will fail the bank that adopts it.

SCALE AND RESOURCES

The most obvious difference is scale. Community banks typically lack dedicated data science teams, AI specialists, or the budget to build custom models. Most AI capabilities come through vendor relationships rather than internal development. That is a structural reality, and governance programs must accommodate it. A governance program that assumes in-house model development capacity will fail at most community banks. The program in this book is designed around the reality that community banks are primarily AI consumers.

RELATIONSHIP-BASED CULTURE

Community banks exist because of relationships. The banker who knows a borrower's character, the branch manager who understands a local business's seasonal cash flow patterns, the board member who lives in the community the bank serves—these relationships are the community banking model. AI governance must protect them.

This means, for example, that an AI system that automates lending decisions must be governed in a way that preserves human judgment in the process, maintains the bank's ability to explain decisions to borrowers in personal terms, and avoids treating customers as data points rather than neighbors. It also means that AI-driven efficiency gains should free bankers to spend more time on relationships.

REGULATORY PROPORTIONALITY

Community banks are subject to the same fundamental regulatory expectations as large banks—fair lending, safety and soundness, consumer protection—but regulators generally apply these expectations with proportionality. A community bank is not expected to maintain the same model risk management infrastructure as a globally significant institution. However, proportionality is not a pass. Regulators increasingly expect community banks to demonstrate that they understand the AI systems they use, that they have assessed the risks, and that they have implemented appropriate controls. "We didn't know our vendor was using AI" is not an acceptable answer.

DECISION-MAKING SPEED

One genuine advantage community banks hold is agility. Without the bureaucratic layers of a large institution, a community bank can design, approve, and implement an AI governance program in weeks rather than months or years. The board, the governance committee, and the IT leader can sit in the same room, make decisions, and move forward. This book is designed to capitalize on that advantage.

COMMUNITY ACCOUNTABILITY

Community bankers live in the communities they serve. If an AI system produces a discriminatory lending outcome, the consequences land on neighbors, businesses, and institutions the banker sees every day. This proximity creates a powerful natural incentive for responsible AI governance that large institutions, with their geographic and organizational distance from affected communities, may lack.

> **Board Perspective**
> Community banks have a governance advantage they rarely recognize: proximity. Your board members live in the community affected by AI decisions. Your IT leader knows the lending team personally. Your governance committee can convene quickly. These are structural strengths that large banks cannot replicate.

1.2 The Cost of Inaction vs. the Cost of Unmanaged Adoption

AI presents community banks with a two-sided risk. The first risk is inaction: failing to adopt AI tools that could improve efficiency, enhance customer service, strengthen compliance, and maintain competitiveness. The second risk is unmanaged adoption: deploying AI without adequate governance, creating exposure to regulatory enforcement, discriminatory outcomes, data breaches, and reputational damage. The enforcement risk has already materialized: in 2025, a state attorney general extracted a $2.5 million settlement from a lender over algorithmic discrimination—a case Chapter 4 returns to.

Both risks are real, and both are growing.

THE COST OF INACTION

Community banks that avoid AI entirely face several escalating pressures.

Competitive erosion. Fintech firms and large banks are putting AI to work where customers can feel it: the fintech down the street approves a small business loan in ten minutes, and the big bank's app explains a customer's spending before the customer asks. A community bank that cannot answer either pace risks losing customers—starting with the younger ones who do all their banking from a phone.

Operational inefficiency. As labor markets tighten and costs rise, banks that do not put automation and AI-driven efficiency to work face structural cost disadvantages. The bank down the road that automates its document processing and compliance reporting can redeploy those staff hours to revenue-generating and relationship-building activities.

Compliance effectiveness. Paradoxically, AI can both create and reduce compliance risk. Banks that fail to adopt AI-powered compliance monitoring tools may find themselves less effective at detecting fraud, suspicious activity, and fair lending violations than their AI-equipped peers—and effectiveness gaps eventually surface in examination results.

Talent attraction. The best young bankers want to work at institutions that invest in technology. A bank perceived as technologically stagnant will struggle to recruit and retain them.

THE COST OF UNMANAGED ADOPTION

The opposite extreme—adopting AI without governance—carries its own serious risks.

Regulatory enforcement. Banking regulators expect banks to govern AI systems with the same rigor applied to other significant risk areas. A bank that cannot explain how its AI lending model makes decisions, or that has never tested a vendor's AI tool for bias, is exposed to supervisory criticism, enforcement actions, and consent orders.

Discriminatory outcomes. AI systems trained on historical data can perpetuate and even amplify existing biases. A lending model that inadvertently produces disparate impact on protected classes can result in fair lending violations, litigation, and community harm—exactly the outcomes community banks exist to prevent.

Data breaches and privacy violations. AI tools, particularly generative AI, create new vectors for data leakage. An employee who uploads customer data to an external AI platform may inadvertently expose that data to unauthorized access, violating GLBA, state privacy laws, and customer trust.

Reputational damage. For a community bank, reputation is everything. A single AI-related incident—a discriminatory lending outcome, a chatbot that gives harmful financial advice, a data breach through an AI tool—can undermine decades of carefully built community trust.

> **The Governance Sweet Spot**
>
> Eliminating AI risk would require eliminating AI—neither possible nor desirable. The goal is to manage AI risk proportionate to its magnitude, the same way banks manage credit risk, operational risk, and compliance risk. Governance is the mechanism that allows a community bank to adopt AI confidently, knowing that risks are identified, measured, monitored, and controlled.

1.3 A Brief Regulatory Landscape

Community bank leaders often ask a straightforward question: "What exactly are the regulations for AI in banking?" The honest answer is that there is no single, comprehensive AI regulation for banks. Instead, AI is governed through a patchwork of existing regulations, supervisory guidance, and emerging rulemaking that collectively create a clear, if complex, set of expectations. Chapter 4 provides a detailed analysis of the regulatory landscape. Here, we provide a high-level orientation.

EXISTING FRAMEWORKS THAT APPLY TO AI

Model Risk Management (SR 26-2 / OCC Bulletin 2026-13, replacing SR 11-7 and OCC Bulletin 2011-12). The revised interagency guidance on model risk management, issued in April 2026, applies directly to machine learning models used in credit decisions, fraud detection, and pricing. SR 26-2 introduces a materiality-based framework allowing banks to calibrate governance effort to actual risk, replaces annual revalidation with risk-triggered reviews, and formalizes proportionality principles that align well with community bank operations. Notably, generative AI and agentic AI are explicitly excluded from its scope—making the broader governance approach in this book essential for institutions using those technologies.

Fair Lending Laws (ECOA, Reg B, Fair Housing Act). These laws prohibit discrimination in lending and apply regardless of whether a human or an algorithm makes the decision. Federal guidance has long emphasized that using AI does not reduce a lender's fair lending obligations—if anything, it increases the need for testing and monitoring. That principle endures even though the Consumer Financial Protection Bureau, which advanced it, sharply curtailed its supervision and enforcement activity in 2025–2026. Fair lending examinations of community banks continue uninterrupted through their prudential regulators—the OCC, the FDIC, and the Federal Reserve—and the obligation also runs through ECOA's private right of action, the Fair Housing Act, state fair lending laws, and state attorneys general, as Chapter 4 explains.

Third-Party Risk Management (OCC 2023-17 / Fed SR 23-4 / FDIC FIL-29-2023, replacing OCC 2013-29 and FIL-44-2008). Most community banks access AI through vendor relationships, which makes existing third-party risk management guidance directly relevant. Banks are expected to conduct due diligence on vendor AI capabilities, understand how AI systems work, and maintain oversight throughout the vendor relationship. This guidance applies equally to open-source AI models, which are increasingly accessible through cloud platforms. Open-source does not mean risk-free: banks still need to validate, document, and monitor these models, and the absence of a commercial vendor relationship shifts more of the governance burden onto the bank itself.

Data Privacy (GLBA, State Laws). The Gramm-Leach-Bliley Act and its implementing regulations require banks to protect customer data—a requirement that extends to any AI system that processes, stores, or transmits such data.

EMERGING REGULATORY DEVELOPMENTS

Beyond existing frameworks, several developments signal the direction of future regulation. Federal banking agencies have folded AI into existing interagency guidance—model risk, third-party risk—and the next round of rulemaking is forming now. State legislatures have begun passing AI-specific laws, some of which may affect banking operations. International frameworks, particularly the EU AI Act, are influencing global standards and may affect US banks with international operations or vendor relationships.

The practical implication for community banks is this: you do not need to wait for a specific AI regulation to begin governing AI. The existing regulatory framework already requires it. Banks that proactively build governance programs today will be well positioned when more specific requirements arrive—and they will be far better prepared for their next regulatory examination.

> **Examiner's Perspective**
> Examiners evaluating a community bank's AI use are looking for evidence that the bank understands what AI systems it uses, has assessed the associated risks, has implemented controls proportionate to those risks, and has appropriate board and management oversight. It is a short list, and every item on it is buildable in a quarter—the Quick Start exists for exactly that.

1.4 Setting the Stage: Governance as Competitive Advantage

There is a natural tendency to view governance as a cost center—an overhead expense imposed by regulators that adds complexity without adding value. This book rejects that framing entirely.

AI governance, done well, is a competitive advantage. It lets a bank adopt AI with confidence, speed, and sustainability. Consider the alternative: a bank without governance must approach every AI decision from scratch, debating risks, seeking ad hoc approvals, and hoping nothing goes wrong. A bank with governance has pre-established criteria for evaluating AI use cases, clear approval pathways, documented risk tolerances, and monitoring systems that catch problems early. The governed bank moves faster.

Governance also creates value in less obvious ways.

Customer trust. Customers who know their bank governs AI responsibly are more likely to trust AI-driven services. Public concern about AI keeps rising, and a community bank that can articulate its AI principles and governance practices has a powerful differentiator.

Regulatory confidence. Good governance buys no immunity—examiners weigh risk, and a well-documented program wrapped around an untested model is still a finding. What it buys is a shorter, better conversation: the bank that can produce its inventory, its policies, and four quarters of committee minutes answers the first AI question in an afternoon. The bank that cannot starts a remediation project under examination pressure. Supervisory relationships are built one afternoon like that at a time.

Employee empowerment. Clear policies and training give employees the confidence to use AI tools effectively. Without governance, employees either avoid AI (losing productivity) or use it recklessly (creating risk). Governance enables responsible adoption.

Strategic clarity. A governance program forces the bank to articulate its AI strategy: which use cases to prioritize, what risk appetite to accept, how to allocate technology investments. This strategic clarity has value far beyond compliance.

One more foundation belongs in this opening chapter: data. AI governance is only as strong as the data underneath it. Every AI system your bank runs—vendor-embedded or homegrown—consumes data that is accurate and current, or data that is fragmented, inconsistent, and stale. The first kind can be governed; the second kind undermines every control you build on top of it, because no policy, committee, or monitoring plan can fix outputs built on broken inputs. That is why the 90-Day Quick Start has you map the data behind each system in its first action, and why Chapter 10 treats data readiness as a governance priority rather than an IT cleanup project. The banks that take data seriously in week one save themselves the most expensive lesson in AI adoption.

How much AI a bank adopts will matter less than how wisely it adopts it. The banks that thrive will pair adoption with clear governance, proportionate controls, and an unwavering commitment to the communities they serve. Think back to the president on that Tuesday morning: none of his three problems required a prediction—everything he needed to know was already on his desk. What he lacked was a frame to put it in and a forum to decide in. This book builds both. We will come back to him at the end.

Chapter Summary

This chapter has established six foundational concepts that underpin the rest of this book:

- AI is already present in community banking—arriving through vendor updates, employee adoption, and competitive pressure—whether or not the bank has made a deliberate decision to adopt it, and the next wave is arriving faster than most governance programs can absorb.
- Community banks have unique characteristics—scale, relationships, agility, and community accountability—that demand a governance approach designed for their context.
- Both inaction and unmanaged adoption carry significant costs; governance is how the bank steers between them.
- The existing regulatory framework already requires AI governance through model risk management, fair lending, third-party risk, and data privacy obligations.
- AI governance is only as strong as the data underneath it. Dirty data produces dirty AI; data readiness is a governance priority.
- Governance is a competitive advantage—it enables faster adoption, builds customer trust, improves regulatory relationships, and provides strategic clarity.

Looking Ahead

Chapter 2 maps the territory this chapter has opened—the specific shapes AI takes in a community bank and where it is already at work. Chapter 3 then establishes the core governance principles—proportionality, transparency, fairness, accountability, privacy, and safety—that will guide every policy, process, and decision tool in this book.

"I asked when we started using AI.
The honest answer was that nobody ever decided to."

— COMMUNITY BANK DIRECTOR —

✦ CHAPTER 2 ✦

The AI You Already Have

WHAT COUNTS, WHERE IT RUNS, AND WHY THERE IS MORE OF IT THAN YOU THINK

Chapter 1 made the case for urgency. This chapter makes the case specific. It defines what counts as AI for governance purposes, then walks through where those systems already operate in a typical community bank—department by department, including the places nobody thinks to look. By the end, you should be able to read your own bank the way an examiner would: as an institution already running on algorithms, whether or not anyone has written that down.

2.1 What Counts as AI in a Community Bank

Before a bank can govern artificial intelligence, it must first define it—and that turns out to be harder than it sounds. The term "artificial intelligence" means different things to different people within the same organization. Board members may think of the AI assistant on their phone or the chatbot on the bank's website. IT leaders may think of the machine learning models running in production. Compliance officers may focus narrowly on the algorithmic models addressed in existing model risk management guidance. Each of these mental models captures a piece of the reality, but none captures the whole.

For the purposes of this book, and for practical governance, we define AI broadly: any system that uses computational techniques to perform tasks that traditionally required human judgment, pattern recognition, or decision-making. The categories below cover most of what a community bank will find.

MACHINE LEARNING MODELS

These are statistical models that learn patterns from data and make predictions or decisions. In community banking, machine learning appears most commonly in credit scoring, fraud detection, anti-money-laundering transaction monitoring, and marketing analytics. Some of these models are built in-house, but the vast majority are embedded in vendor platforms and may operate without anyone at the bank having been told.

ROBOTIC PROCESS AUTOMATION (RPA)

RPA tools automate repetitive, rules-based tasks such as data entry, report generation, account reconciliation, and document processing. While RPA in its simplest form follows rigid rules, modern RPA increasingly incorporates machine learning components that introduce adaptive behavior—blurring the line between automation and AI.

NATURAL LANGUAGE PROCESSING AND GENERATIVE AI

This category includes chatbots, virtual assistants, document summarization tools, and large language model tools like ChatGPT. Generative AI is the most visible form of AI entering community banks, and the most disruptive. Employees may already be using these tools informally, creating shadow AI: AI use that exists outside the bank's knowledge and governance structures.

VENDOR-EMBEDDED AI

The most significant and least understood category in community banking. Core processing platforms, digital banking solutions, loan origination systems, and compliance monitoring tools increasingly embed AI features that are activated by default or through routine software updates. A community bank may be "using AI" without having made a conscious decision to adopt it, simply because a vendor upgraded its platform.

AGENTIC AI

Agentic AI is the category to prepare for now, while deployment is still early enough to prepare on your own terms. Unlike traditional AI that answers a question or generates a recommendation for a human to act on, agentic AI systems can plan, reason, and execute multi-step workflows autonomously—researching a borrower, assembling a credit package, drafting a recommendation, and routing it for approval with minimal human intervention.

Vendors are already beginning to embed agentic capabilities into core banking platforms, compliance monitoring tools, and customer service systems. The first enterprise agents are also arriving inside the office productivity platforms banks already run—with their own identities and standing access—which makes identity and access management part of AI governance. The governance challenge is fundamental: agentic AI reduces the number of points at which a human can intervene, which means the controls, guardrails, and kill switches must be designed into the system before it is deployed. Chapter 11 addresses agentic AI governance in detail, and Chapter 17 surveys what is coming next.

2.2 Where AI Already Lives in Your Bank

The fastest way to make AI governance concrete is to tour the bank. Every system below is drawn from platforms community banks already run; nothing here is speculative. Chapter 15 examines these use cases in governance depth—the purpose of this tour is recognition.

LENDING AND CREDIT

Credit bureau scores are model outputs, and most loan origination systems now layer additional scoring on top of them. Automated valuation models estimate collateral values. Document-extraction features read tax returns and pay stubs into the underwriting file. If the bank participates in indirect lending, the dealer platform's instant decisioning is a model rendering credit judgments under your name. Pricing engines and prequalification widgets on the bank's website round out the lending layer.

FRAUD, BSA/AML, AND PAYMENTS

This is the densest AI territory in any community bank. Debit and credit card fraud scoring runs at the processor on every transaction. The AML system flags unusual activity using rules increasingly supplemented by machine learning. Check fraud detection, wire screening, and sanctions list matching all rely on algorithmic judgment—fuzzy name matching is a model decision about how close is close enough. Most of these systems arrived through the core or processor relationship, and most have been running for years.

DEPOSIT OPERATIONS AND THE BACK OFFICE

Robotic process automation reconciles accounts and moves data between systems that refuse to talk to each other. Optical character recognition reads deposited checks and new-account documents. Loan servicing platforms route exceptions, collections software prioritizes outreach, and report automation assembles the board package. Individually mundane, collectively a substantial automated workforce.

CUSTOMER CHANNELS

The website chatbot and the phone system's speech recognition are customer-facing AI in the most literal sense. Inside mobile banking, transaction categorization, cash-flow insights, and savings nudges are model-driven personalization. The marketing platform's next-best-offer engine decides which customers see which products—a quiet algorithm with fair lending implications Section 15.6 takes up directly.

THE EMPLOYEE LAYER

Email filtering and phishing detection have been machine learning for a decade. Newer arrivals carry more governance weight: the AI assistants embedded in office productivity suites, meeting transcription tools, browser extensions, and the public generative AI tools employees reach for on their own initiative. That last category is the subject of the next section.

Every system in this tour shares one trait: it is visible somewhere—in a contract, an invoice, or an admin console. The next category leaves no paper trail at all.

2.3 Shadow AI and the Unmanaged Footprint

Shadow AI is one of the most immediate and least managed risks facing community banks. When a loan officer pastes customer financial data into a free AI tool to help draft a credit memo, that action creates data privacy risk, regulatory risk, and reputational risk—all outside the bank's governance program.

Shadow AI is almost always well-intentioned. Employees turn to AI tools because those tools are genuinely helpful—they save time, improve quality, and reduce tedium. The governance challenge is to channel that impulse: provide approved tools, clear boundaries, and the training to understand why those boundaries exist. Increasingly, AI also arrives through automation platforms—process-discovery tools that record how employees work to find automation candidates, and no-code builders that let staff create their own automations. Both belong in the bank's AI inventory and intake process, whether or not any single component is technically "AI."

Channeling has three working parts: an approved tool good enough that employees stop reaching for the free one; data boundaries stated in plain language and repeated until they are reflex; and an intake path that lets an employee propose a new AI use without assuming the answer will be no. Banks that supply all three watch shadow AI shrink on its own. Banks that supply none of them get a policy that is obeyed on the survey and ignored at the desk.

2.4 Why You Have More AI Than You Think

The arithmetic of vendor-embedded AI explains why nearly every inventory comes back larger than expected. A community bank typically runs twenty to forty significant systems—core processing, digital banking, loan origination, card processing, AML, HR, marketing. Each vendor adds AI features on its own release schedule, frequently enabled by default, and announces them in release notes few banks read closely.

The result is adoption by upgrade. A bank that has "not adopted AI" discovers it activated AI-driven features in three platforms during routine maintenance windows—with no risk assessment, no contract addendum, and no board mention. There was no decision point, so no governance process was triggered. The bank's AI footprint grew on a vendor's timetable while its governance stood still.

This is why the AI inventory in Chapter 8 is the first concrete action of the governance program, sequenced before the policies are finalized: you write better rules once you can see what they must cover. Appendix F provides the inventory worksheet, and the Quick Start schedules the work into the program's first two weeks.

Walk this tour in your own hallways and the abstraction disappears: the bank has been adopting AI for years. The rest of this book is about governing what the tour reveals.

> **Key Takeaway for Governance**
>
> Your bank is using AI today; the open question is how much. Effective governance begins with a comprehensive inventory of all AI systems, including those embedded in vendor products. Chapter 8 provides a detailed methodology for building this inventory.

Chapter Summary

This chapter has mapped what AI is and where it already operates:

- For governance purposes, AI is any system performing tasks that once required human judgment—machine learning models, RPA, generative AI, vendor-embedded features, and agentic systems.
- A function-by-function tour of a typical community bank surfaces AI in lending, fraud and BSA/AML, deposit operations, customer channels, and the employee toolset—most of it acquired through vendors.
- Shadow AI—employee use of AI tools outside the bank's knowledge—is among the most immediate unmanaged risks, and it is almost always well-intentioned.
- Vendor release cycles produce adoption by upgrade: AI features arrive enabled by default, with no decision point to trigger governance.
- The remedy begins with visibility—the AI inventory of Chapter 8, structured by the worksheet in Appendix F.

Looking Ahead

Chapter 3 establishes the six core governance principles—proportionality, transparency, fairness, accountability, privacy, and safety—that turn this map into a program. Every part of the program in the rest of the book traces back to them.

Standing Up the Inventory

The AI Inventory and Classification Worksheet in Appendix F gives you the structure. The AI Governance Toolkit gives you the editable Excel workbook—with dropdown menus, filtering, and pre-built fields—so your team can stand up the inventory in a single working session and fill it as vendor responses arrive. See the Toolkit page at the back of this book.

"The technology changes every quarter. What we believe about serving this community hasn't changed in forty years. Start there."

— COMMUNITY BANK PRESIDENT —

✦ CHAPTER 3 ✦

Governing AI: The Six Principles

SIX PRINCIPLES THAT DECIDE THE HARD CASES

Chapter 1 established that AI is already present in community banking and that governance is both a regulatory expectation and a competitive advantage. This chapter answers the next logical question: What principles should guide that governance?

Principles matter because they provide a decision-making foundation that extends beyond any single policy or procedure. Policies will change as technology evolves and regulations shift. Principles endure. When a community bank encounters an AI governance question not addressed by its existing policies—and this will happen frequently—these principles decide the case.

This chapter establishes six core principles, each grounded in the practical realities of community banking and aligned with emerging regulatory expectations. Together, they are the working assumptions behind every policy, process, and decision tool in this book.

PRINCIPLE	CORE QUESTION IT ANSWERS	PRIMARY AUDIENCES
Proportionality	How much governance is enough?	All
Transparency	Can we explain AI decisions?	IT Leaders, Boards
Fairness	Are AI outcomes equitable?	Governance Committees, IT Leaders
Accountability	Who owns the AI and its outcomes?	All
Privacy	Is customer data protected?	IT Leaders, Governance Committees
Safety & Resilience	What happens when AI fails?	IT Leaders, Boards

3.1 Proportionality: Right-Sizing Governance to Risk

Proportionality is the most important principle in this book—the difference between governance that works and governance that collapses under its own weight.

The concept is straightforward: the intensity of governance should match the intensity of risk. A simple RPA tool that automates data entry from one internal system to another does not require the same governance rigor as a machine learning model that makes consumer lending decisions. A chatbot that answers frequently asked questions about branch hours poses different risks than a generative AI tool that drafts customer communications. Treating all AI systems identically—either with maximum governance or with none at all—is a recipe for failure.

WHY COMMUNITY BANKS MUST GET PROPORTIONALITY RIGHT

Large banks can afford to over-govern. They have teams of model risk professionals, dedicated AI ethics officers, and budgets that can absorb the cost of extensive governance processes applied uniformly across all AI systems. Community banks cannot. If every AI system requires a 40-page risk assessment, a quarterly validation review, and a board presentation, the governance program will either collapse from resource exhaustion or prevent the bank from adopting AI at all.

The solution is a tiered approach to governance. This book uses a four-tier risk classification system, detailed in Chapter 8, that categorizes AI systems as Minimal, Moderate, High, or Critical risk based on objective criteria. Governance requirements—documentation, validation, monitoring, approval authority, and reporting—scale with the tier. A minimal-risk system might require only a brief inventory entry and periodic review. A critical-risk system requires comprehensive documentation, independent validation, continuous monitoring, and board-level oversight.

> **Proportionality in Practice**
> A $1 billion community bank needs governance that is deliberate, documented, and appropriate to its AI risk profile—a fraction of a $500 billion institution's infrastructure, and just as real. Proportionality is the discipline of doing the right things at the right scale.

THE PROPORTIONALITY CALIBRATION QUESTIONS

When assessing how much governance an AI system requires, the governance committee should consider five calibration questions:

1. **Impact:** What is the potential harm if this AI system produces an incorrect or biased outcome? Could customers be denied credit, flagged as suspicious without cause, or receive inappropriate advice?

2. **Autonomy:** Does this AI system make decisions autonomously, or does it assist a human decision-maker? Systems that act independently require more governance than those that inform human judgment.

3. **Regulatory sensitivity:** Does this AI system touch areas of heightened regulatory scrutiny, such as fair lending, BSA/AML, or consumer protection?

4. **Data sensitivity:** What type of data does this AI system process? Systems that use personally identifiable information, financial data, or protected class data require elevated governance.

5. **Complexity:** How complex and opaque is the AI system? A simple rules-based model is easier to govern than a deep learning model whose decision logic cannot be readily interpreted.

In Chapter 8, these five questions become the formal scoring dimensions of the risk classification matrix—the same lens, sharpened into a tool.

3.2 Transparency: Explainability and Disclosure

If a community bank cannot explain how an AI system reaches its conclusions, it cannot meaningfully govern that system. Transparency—the ability to understand and articulate what an AI system does and why—underpins every other governance activity: risk assessment, bias testing, regulatory compliance, customer communication, and board oversight.

Explainability is audience-specific: the regulator, the customer, and the board each need a different answer to the same question—how does this system work?

EXPLAINABILITY FOR REGULATORS

Examiners expect banks to demonstrate a substantive understanding of the AI systems they use—articulated in clear language: what data the AI system uses, what decisions or recommendations it produces, how its performance is measured, and what controls are in place to manage its risks. No one expects your lenders to explain gradient descent. For vendor-supplied AI, this means obtaining sufficient documentation from vendors to support these explanations.

EXPLAINABILITY FOR CUSTOMERS

Customers have a right to understand the basis for decisions that affect them—an ethical principle that is also, in many contexts, a legal requirement. Adverse action notices under the Equal Credit Opportunity Act (ECOA) must identify the principal reasons for a credit denial. When an AI model drives the denial, the bank must be able to translate the model's factors into the specific, meaningful reasons the law requires. "The computer said no" is never an acceptable explanation.

EXPLAINABILITY FOR THE BOARD

The board needs the shortest version and the highest stakes. Board-level AI reporting should translate technical AI performance metrics into business risk language: What decisions is this AI making? How do we know it is making them correctly? What could go wrong, and what are we doing about it? Chapter 13 provides detailed guidance on board-level AI reporting.

THE VENDOR TRANSPARENCY CHALLENGE

The most significant transparency challenge for community banks involves vendor-supplied AI. Vendors may claim proprietary protection over their algorithms, making it difficult for banks to understand how AI decisions are made. The business logic is understandable; the governance gap it leaves is unacceptable. Community banks must negotiate for adequate transparency in vendor contracts—sufficient documentation of model inputs, logic, performance metrics, and testing results to support meaningful oversight. Chapter 9 provides specific contractual language and negotiation strategies.

> **IT Leader's Note**
> When a vendor tells you their AI model is a "black box," that is a risk management finding to pursue. What you need is the inputs, the outputs, the performance metrics, and the testing methodology—source code access is rarely the point. If a vendor cannot or will not provide this, weigh it heavily in your vendor selection decision.

3.3 Fairness: Bias and Fair Lending Obligations

Fairness is a legal obligation for banks, and an ethical commitment that runs deeper than the law. AI systems, despite their mathematical precision, can produce unfair outcomes—and often do. This happens because AI learns from historical data that may reflect decades of human bias, structural inequality, and discriminatory patterns—and it reproduces what it learns.

Consider a lending model trained on historical loan performance data. If, historically, certain geographic areas or demographic groups received fewer loans—because of discriminatory practices rather than creditworthiness—the model may learn to associate those characteristics with higher risk. The model carries no intent, and the outcome is discriminatory all the same: certain groups face systematically worse treatment.

FAIR LENDING IN THE AI CONTEXT

The Equal Credit Opportunity Act prohibits discrimination on the basis of race, color, religion, national origin, sex, marital status, age, receipt of public assistance, or the good-faith exercise of rights under the Consumer Credit

Protection Act. The Fair Housing Act extends similar protections to residential lending. These laws apply with full force to AI-driven decisions. A bank cannot defend a discriminatory outcome by pointing to an algorithm—the bank remains responsible for the result regardless of the method used to produce it.

For community banks, this means that any AI system involved in lending decisions—whether a credit scoring model, a pricing algorithm, or an automated pre-screening tool—must be tested for disparate impact on protected classes. That obligation survives the 2026 Regulation B changes: the Fair Housing Act, state fair lending laws, and basic safety and soundness all still demand it. This testing is not optional, and it must be ongoing.

BEYOND LENDING: FAIRNESS ACROSS AI APPLICATIONS

Fairness reaches past the loan file. When the marketing engine steers product offers away from certain demographics, the bank owns a fair lending and fair banking problem—whatever the engine intended. When the chatbot gives some customers worse answers than others, the bank carries the reputational and compliance fallout. And the obligation follows AI inside the building: AI-assisted hiring and performance evaluation answer to employment law's fairness requirements.

PRACTICAL APPROACHES TO BIAS MANAGEMENT

Chapter 5 provides detailed guidance on bias detection and mitigation. At the principle level, community banks should commit to three practices:

1. **Test before deployment.** Every AI system that affects customers should undergo bias testing before it goes live, using appropriate statistical methods for the type of decision being made.

2. **Monitor continuously.** Bias can emerge or shift over time as data distributions change. A clean test at deployment is a snapshot, and snapshots age.

3. **Act on findings.** Identifying bias is meaningless without a commitment to address it. The governance program must include clear escalation paths and remediation procedures when bias is detected.

3.4 Accountability: Ownership and Oversight

Every AI system in a community bank must have a named human owner.

In many organizations, AI accountability falls into a gap. The vendor built the model, so the vendor is responsible. The IT department deployed it, so IT is responsible. The business line uses it, so the business line is responsible. The result is that no one is truly responsible, and when something goes wrong, accountability is diffused to the point of meaninglessness.

The accountability principle requires that every AI system have a designated owner—typically a senior manager in the business line that benefits from the system—who is personally responsible for ensuring that the system is governed appropriately. The owner need not perform every governance activity personally; the owner is the single point of accountability for ensuring that governance activities occur, that risks are managed, and that issues are escalated.

THE ACCOUNTABILITY CHAIN

Accountability in AI governance operates at three levels:

1. **System-level ownership:** A named individual who owns each AI system's governance, including documentation, risk assessment, monitoring, and issue resolution.

2. **Committee-level oversight:** The AI governance committee (or equivalent body) that provides collective oversight across all AI

systems, challenges management assumptions, and approves high-risk deployments.

3. **Board-level responsibility:** The board of directors, which bears ultimate fiduciary responsibility for the bank's AI risk posture and ensures that management has established adequate governance.

Chapter 6 maps these accountability relationships in detail using a RACI matrix that assigns specific responsibilities to IT, compliance, risk, business lines, the governance committee, and the board.

> **Board Perspective**
>
> The board owns the governance program; management owns the systems. The board's role is to ensure that management has established an effective AI governance structure, that it is resourced appropriately, and that it is functioning as intended. Chapter 13 provides specific guidance on how boards can fulfill this responsibility without micromanaging.

3.5 Privacy: Data Protection and Customer Trust

AI runs on data, and in banking that data is deeply personal: income, debts, spending patterns, employment history, family status, health-related transactions, and more. The community banking relationship is built on trust, and customers trust that their bank will protect this information with the same care a physician applies to medical records.

AI introduces new privacy challenges that go beyond traditional data security. When an AI model is trained on customer data, that data's influence persists in the model's learned patterns even after the original data is deleted. When an employee enters customer information into a generative AI tool, that data may be processed on external servers, stored in ways the bank cannot control, and potentially exposed to unauthorized access. When a vendor's AI platform analyzes customer behavior to generate marketing insights, the boundary between helpful personalization and invasive surveillance can be difficult to define.

DATA GOVERNANCE AS AI GOVERNANCE

Effective AI governance is inseparable from effective data governance. Before a community bank can govern its AI systems, it must govern its data: understanding what data it collects, where that data is stored, how it flows through AI systems, who has access, and what controls protect it. Chapter 7 addresses data governance policy specifically, and Chapter 10 (Section 10.2) provides IT leaders with practical guidance on data readiness.

THE ETHICS OF DATA USE

Legal compliance is the floor. Community banks should aspire to data practices that go beyond what the law requires, guided by a simple question: Would our customers be comfortable if they knew how we were using their data? If the answer is uncertain, the practice warrants another look.

This ethical lens is particularly important for community banks because the cost of lost trust is existential. A large national bank can absorb a data controversy with marketing spend and institutional inertia. A community bank's reputation is its franchise. Once lost, customer trust is extraordinarily difficult to rebuild.

3.6 Safety & Resilience: Failure Modes and Recovery

AI systems will fail. Models will produce incorrect predictions. Chatbots will generate inappropriate responses. Fraud detection algorithms will miss fraudulent transactions or flag legitimate ones. Data feeds will be corrupted, software updates will introduce bugs, and vendor platforms will experience outages. The governance task is to be ready when it happens.

The safety principle requires that AI systems be designed, deployed, and governed with failure in mind. That means safeguards that keep failures from causing serious harm, monitoring that catches them quickly, human overrides that let staff intervene, and incident response procedures that make remediation fast.

HUMAN OVERRIDE AND ESCALATION

No AI system in a community bank should operate without the possibility of human intervention. A human need not approve every AI decision—that would negate the efficiency benefits of AI. What the principle requires is a clear, practiced mechanism for a human to override the AI's recommendation and escalate the matter for review whenever an outcome appears anomalous, unjust, or harmful.

The design of human override mechanisms requires careful thought. An override that exists on paper and is never used protects nothing. Overrides must be accessible, practical, and supported by a culture that empowers employees to exercise judgment even when it contradicts the AI's output.

The Agentic AI Challenge

Agentic AI raises the stakes on every safety principle in this section. Traditional AI systems make a recommendation and wait for a human to act on it—a loan officer reviews the credit score; a compliance analyst reviews the alert. Agentic AI systems run the same multi-step workflows Chapter 2 described—research to recommendation to routing—with minimal human touchpoints. Each step the agent takes without human review is a step where errors compound, biases propagate, and controls cannot intervene. Chapter 11 builds the full governance model for agentic AI.

For community banks, this means that the governance guardrails—kill switches, human escalation triggers, output validation checkpoints, and activity logging—must be designed into agentic systems before deployment. Your bank will encounter agentic AI through the platforms you use today. The open question is whether your governance program is ready for the moment the AI stops asking for permission and starts acting on its own.

RESILIENCE AND BUSINESS CONTINUITY

AI resilience intersects with traditional business continuity planning. If a critical AI system goes down—the fraud detection model, the automated lending platform, the chatbot that handles after-hours customer inquiries—what is the fallback? Community banks must ensure that human-driven backup processes exist for every critical AI system and that staff are trained to execute those processes under stress.

> **Key Takeaway**
> Safety means ensuring that imperfect AI cannot cause disproportionate harm. The combination of monitoring, human override, incident response, and business continuity planning creates layers of protection that allow a community bank to benefit from AI while managing the inevitable failures.

Six principles, and none of them new. Fairness, accountability, prudence, and care for the customer have governed community banking since before anyone called them principles. You already have what you need to govern AI—you've been practicing the principles for decades. This book just shows you how to point them at something new.

Chapter Summary

This chapter has established six core governance principles that form the foundation for every policy, process, and tool in this book:

- Proportionality ensures governance effort matches risk—neither too much nor too little.
- Transparency requires that AI systems be explainable to regulators, customers, and the board.

- Fairness demands active testing and monitoring for bias, particularly in lending and customer-facing applications.
- Accountability assigns named human ownership for every AI system and establishes a clear chain of responsibility.
- Privacy extends data protection obligations to AI-specific risks and grounds data use in customer trust.
- Safety prepares the bank for AI failures through monitoring, human override, incident response, and resilience planning.

Looking Ahead

Chapter 4 maps the regulatory landscape in detail, tying specific regulatory requirements to the governance actions community banks must take. Chapter 5 then translates the fairness and ethics principles into operational practice. With principles established and regulations understood, Part III of the book turns to the practical work of designing and building the governance program itself.

From Principles to Policy

Appendix A translates these principles into a complete governance policy. The AI Governance Toolkit provides it as an editable Word document—complete the bracketed fields with your bank's name, titles, and structure, and it's ready for board review.

PART II

The Rules That Already Apply

"What regulators need from you is thoughtfulness, documentation, and honesty about what you don't know. Perfection has never been the standard."

— FORMER OCC EXAMINER —

✦ CHAPTER 4 ✦

The Regulatory Landscape

A DETAILED MAP OF CURRENT AND EMERGING REQUIREMENTS

AI in banking answers to layers of authority—statutes, regulations, supervisory guidance, examination practice—and together they set the expectations for how your bank develops, deploys, and oversees every AI system it runs. For a community bank leader already managing a heavy regulatory burden, the encouraging news is structural: almost none of this is new law. It is existing law reaching new tools—through disciplines your bank already practices.

This chapter provides a practical map of the regulatory landscape as it applies to AI in community banking. The goal is working knowledge: enough to make informed governance decisions, engage productively with examiners, and build a compliance program that is defensible and sustainable.

We begin with a reference table mapping key regulations to their AI relevance, then explore each major regulatory domain in detail.

REGULATION / GUIDANCE	AI RELEVANCE	KEY GOVERNANCE ACTION
SR 26-2 (formerly SR 11-7) / OCC Bulletin 2026-13	ML models in credit, fraud, pricing, ALM	Model inventory, validation, documentation, board reporting
ECOA / Reg B	Any AI in credit decisions or pricing	Adverse action notices, disparate-treatment testing, fair lending analysis
Fair Housing Act	AI in residential mortgage decisions	Disparate-impact testing, redlining analysis, marketing review
Bank Secrecy Act (BSA) / Anti-Money Laundering (AML)	AI transaction monitoring, SAR filing	Model validation, tuning documentation, false positive analysis
OCC 2023-17 (formerly 2013-29) / Fed SR 23-4 / FDIC FIL-29-2023	All vendor-supplied AI	Vendor due diligence, contract provisions, ongoing monitoring
GLBA / Reg P	AI processing customer data	Data governance, access controls, privacy notices, data flow mapping
UDAP / UDAAP	Customer-facing AI (chatbots, recommendations)	Output monitoring, disclosure, complaint analysis

4.1 Model Risk Management (SR 26-2 / OCC Bulletin 2026-13)

In April 2026, the Federal Reserve, OCC, and FDIC jointly issued revised model risk management guidance—designated SR 26-2 by the Federal Reserve and Bulletin 2026-13 by the OCC—replacing the original SR 11-7 guidance that had governed model risk management since 2011. The revised guidance modernizes the framework with a principles-based, risk-proportionate approach while preserving the foundational requirements that made SR 11-7 effective. For community banks, the shift is significant: SR 26-2 formalizes materiality-based oversight, replaces the default annual revalidation cycle with risk-triggered reviews, and allows greater flexibility in how validation independence is achieved. It also explicitly excludes generative AI and agentic AI from its scope—making institution-level governance programs like the one in this book essential for covering the full spectrum of AI risk.

The agencies' deregulatory turn reached community banks directly. In OCC Bulletin 2025-26, "Model Risk Management: Clarification for Community Banks" (October 6, 2025), the OCC underscored that a community bank—an institution with up to $30 billion in assets—has the flexibility to tailor its model risk management practices, including the frequency and scope of model validation, to its own risk exposures, business activities, and the complexity and extent of its model use. Critically, the OCC clarified that its guidance does not require community banks to perform annual model validation, and that examiners will not give negative supervisory feedback solely because of the validation frequency or scope a bank reasonably determined for itself. For an institution that buys rather than builds nearly all of its models, this is the clearest signal yet that proportionality is the expected approach—provided the bank can show its choices were deliberate and grounded in risk.

WHAT IS A MODEL UNDER SR 26-2?

SR 26-2 defines a model as a complex quantitative method, system, or approach that applies statistical, economic, financial, or mathematical theories, techniques, and assumptions to process input data into quantitative estimates. The revised definition adds the qualifier "complex" and explicitly excludes simple arithmetic calculations found in spreadsheets, as well as deterministic rule-based processes—narrowing the scope from SR 11-7's famously broad definition. This definition still encompasses virtually all machine learning systems used in banking, including credit scoring models, fraud detection algorithms, interest rate risk models, and customer segmentation tools.

However, SR 26-2 resolves some gray areas that SR 11-7 left open. A simple robotic process automation (RPA) bot that follows rigid, deterministic rules is now explicitly outside the model definition. Generative AI and agentic AI tools are excluded from SR 26-2's scope entirely—though this does not mean they require no governance, only that the guidance does not prescribe how to govern them. Community banks should still err on the side of inclusion: if a system produces outputs that influence business decisions, treat it as subject to sound governance principles, even if SR 26-2 does not technically require it. The governance programs in this book are designed to cover that broader landscape.

THE THREE PILLARS OF MODEL RISK MANAGEMENT

SR 26-2 preserves the three foundational requirements from SR 11-7 that apply directly to AI governance, while introducing materiality-based calibration to each:

1. **Model development and implementation.** Models must be developed using sound theory and methodology, with clear documentation of design choices, assumptions, data inputs, and limitations. For AI systems, this means documenting the training data, the algorithm used, the feature selection process, and the performance metrics—whether the model is built in-house or acquired from a vendor.

2. **Model validation.** Models must undergo independent validation to confirm they perform as intended. SR 26-2 replaces the presumption of annual revalidation with a risk-based approach: validation frequency is now tied to the model's materiality, change velocity, and data availability, with explicit triggers for re-review. The guidance also clarifies that validation quality depends on the rigor and effectiveness of the review rather than on organizational structure alone—a meaningful flexibility for community banks with limited staffing. For complex AI models, validation still requires specialized expertise, and banks that lack in-house capability should engage qualified third parties.

3. **Governance, policies, and controls.** The bank must have policies governing model risk, an inventory of all models, clear roles and responsibilities, and board-level oversight. This is precisely the governance program this book helps you build.

> **Practical Guidance**
>
> For community banks, the most common model risk management gap is the absence of a model inventory, since the models themselves are already everywhere. Many banks use vendor-supplied AI models without recognizing them as models subject to model risk management. The first step is always the inventory: a bank cannot manage models it has never catalogued. See Chapter 8 for the complete inventory methodology.

AI-SPECIFIC CHALLENGES BEYOND SR 26-2

While SR 26-2 modernizes the model risk management framework, AI systems—particularly generative and agentic AI—continue to present challenges that the guidance does not fully address. SR 26-2 explicitly excludes generative AI and agentic AI from its scope, leaving institutions to develop their own governance approaches for these technologies. Even for machine learning models that remain within scope, challenges persist: they can be difficult to interpret, making traditional documentation and validation approaches insufficient. Models that learn continuously from new data may drift from their validated state without any deliberate change. And vendor-supplied AI models may come with limited documentation, making it difficult for banks to fulfill their own model risk management obligations.

These challenges leave community banks with a sharper obligation: governance programs adapted to AI's unique characteristics. This book's risk assessment methodology (Chapter 8) and vendor governance program (Chapter 9) are specifically designed to address these AI-specific gaps.

One sector-specific resource now fills part of that gap: the Treasury Financial Services AI Risk Management Framework, developed by the Cyber Risk Institute and the Financial Services Sector Coordinating Council and released by the U.S. Treasury in February 2026. The framework contains roughly 230 control objectives organized across governance, risk management, and compliance—the most comprehensive sector framework targeting AI specifically in financial services—industry-developed and released through Treasury, so it guides examiners without binding them. For community banks, it matters because it provides concrete control expectations that go well beyond SR 26-2's principles-based approach. You do not need to implement all 230 controls—but you should know what examiners are reading, and this framework will shape examination expectations for years to come.

4.2 Fair Lending and Anti-Discrimination

Fair lending is the highest-stakes regulatory domain for AI in banking. The potential for AI to produce discriminatory outcomes—even unintentionally—creates exposure that is legal, financial, and, for a community bank, intensely local.

THE LEGAL FRAMEWORK

Three statutes anchor this domain: the Equal Credit Opportunity Act and the Fair Housing Act prohibit discrimination in credit decisions, and the Community Reinvestment Act requires banks to affirmatively serve their communities, including low- and moderate-income areas. None of these statutes mentions AI. All of them reach it.

Two theories of discrimination are relevant to AI. Disparate treatment occurs when a lender intentionally considers a prohibited factor in a credit decision. Disparate impact occurs when a facially neutral practice disproportionately affects a protected class without a legitimate business justification. As discussed below, the CFPB finalized the removal of the disparate impact standard from Reg B in 2026—the effects-test language lived in the regulation's commentary. The Fair Housing Act's disparate impact standard still applies to mortgage lending, however, and multiple states have their own disparate impact provisions. The effects test also originated in case law under ECOA itself; courts are not bound by a deletion from the commentary.

FAIR LENDING REGULATION OF AI: FEDERAL AND STATE LANDSCAPE

The federal approach to AI in lending has shifted. In 2022, the CFPB issued Circular 2022-03, which set clear expectations for adverse action notices when complex algorithms are involved in credit decisions—emphasizing that lenders could not use AI complexity as a reason to provide vague or generic denial notices. That circular was withdrawn in May 2025, and in April 2026 the CFPB finalized a rule removing the disparate impact standard from Reg B, effective July 21, 2026.

What has not changed is the legal foundation: ECOA still requires specific, accurate adverse action notices, the Fair Housing Act's disparate impact standard still applies to mortgage lending, and state fair lending laws—which increasingly include their own disparate impact provisions—remain fully enforceable. Community banks that built their adverse action notice processes and fair lending controls around those principles should not dismantle them simply because the federal posture has changed—the legal exposure that justified those controls in the first place remains.

What has changed is who enforces these obligations. Through 2025 and into 2026, the Consumer Financial Protection Bureau—long the most active federal voice on AI in consumer lending—saw its leadership, funding, and enforcement capacity sharply reduced; it withdrew prior guidance and stepped back from supervision and rulemaking. The Bureau, however, was never the fair lending supervisor for banks under $10 billion in assets: community banks have always received their fair lending examinations from their prudential regulator—the OCC, the FDIC, or the Federal Reserve—and that supervision continues exactly as before. ECOA also still directs an agency that finds reason to believe a pattern or practice of discrimination to refer the matter to the Department of Justice. For community banks the expectations remain in force; the enforcement has shifted: the Fair Housing Act (enforced by HUD and the Department of Justice), state fair lending and consumer-protection laws, and increasingly assertive state attorneys general now carry much of the load the Bureau's enforcement once did. Guidance the Bureau has withdrawn still describes sound practice—and a bank that dismantles its controls because one federal enforcer has gone quiet is misreading where the risk now lives.

For community banks, the practical implication is clear: if you use AI in any aspect of lending—origination, pricing, servicing, collections, or marketing—you must be able to demonstrate that the system has been tested for fair lending compliance and that appropriate controls are in place.

State enforcement adds another dimension to this landscape. As federal enforcement priorities have shifted, state attorneys general have stepped into a more active role. In July 2025, Massachusetts reached a $2.5 million settlement with a fintech lender over algorithmic lending discrimination—one of the first state enforcement actions targeting AI-driven credit decisions. The activity has broadened since: state attorneys general are using their consumer-protection statutes—which carry per-violation penalties and require no proof of individual damages—to pursue AI bias and disclosure claims, and a new wave of state AI laws that took effect in January 2026 gives several of them AI-specific enforcement authority. Most community banks operate under state charters and are directly subject to these laws, which makes strong fair lending controls a matter of both regulatory compliance and practical risk management.

GOVERNANCE ACTIONS FOR FAIR LENDING COMPLIANCE

Pre-deployment testing: Conduct disparate impact analysis before any AI lending model goes live.
Ongoing monitoring: Regularly analyze AI lending outcomes by protected class using statistical methods appropriate to the bank's volume and portfolio.
Adverse action procedures: Ensure that AI-driven credit decisions produce specific, meaningful adverse action reasons that comply with Regulation B.
Documentation: Maintain documentation of fair lending testing methodology, results, and any remediation actions taken.
Vendor accountability: Require vendors to provide fair lending testing results and cooperate with the bank's own testing.

Nothing on this list is new discipline—fair lending programs have run these motions for decades. AI raises the stakes and the sample sizes.

4.3 BSA/AML and AI-Driven Transaction Monitoring

BSA/AML is where most community banks have run AI the longest. The monitoring system reads every transaction—a volume no compliance team could review by hand—and surfaces the patterns worth a human look. Tuned well, it catches more suspicious activity and hands your analysts a shorter, smarter alert queue.

TUNING AND THRESHOLD MANAGEMENT

AI transaction monitoring models require ongoing tuning to balance detection effectiveness with false positive rates. An overly sensitive model generates excessive alerts, overwhelming compliance staff and creating the risk that genuine suspicious activity is lost in the noise. An overly permissive model misses suspicious transactions, exposing the bank to regulatory and criminal liability. Governance must include documented tuning procedures, periodic effectiveness reviews, and clear accountability for tuning decisions. The agencies addressed these systems directly in the April 2021 Interagency Statement on Model Risk Management for Bank Systems Supporting BSA/AML Compliance: model risk principles apply to BSA/AML tools flexibly, threshold and scenario updates do not automatically demand full revalidation, and model risk process should never delay a compliance fix the bank knows it needs.

SUSPICIOUS ACTIVITY REPORTING

While AI can flag transactions for review, the decision to file a Suspicious Activity Report (SAR) remains a human responsibility. AI should inform the SAR decision; the judgment to file stays with a person. Banks must ensure that compliance staff have the training and authority to exercise independent judgment when reviewing AI-generated alerts, including the ability to override the model's assessment when circumstances warrant.

Examination Focus Area

Examiners reviewing AI-driven BSA/AML systems focus on three questions: Does the bank understand how the model identifies suspicious activity? Can the bank demonstrate that the model's tuning parameters are appropriate? Is there adequate human oversight of the model's alert output? The three cascade: a bank that cannot answer the first has no basis for the other two. Answer all three confidently and the conversation gets shorter.

4.4 Vendor Management and Third-Party Risk (OCC 2023-17 / Fed SR 23-4 / FDIC FIL-29-2023)

For most community banks, AI governance is, in large part, vendor governance. The majority of AI capabilities come through vendor relationships—core processing platforms, digital banking solutions, fraud detection systems, and compliance monitoring tools. Existing third-party risk management guidance from the OCC, FDIC, and Federal Reserve applies directly to these relationships.

REGULATORY EXPECTATIONS

Banking agencies have consistently emphasized that a bank cannot outsource responsibility by outsourcing a function. When a vendor's AI system produces a biased lending decision, the bank faces the regulatory consequences. This principle applies regardless of the bank's ability to inspect the vendor's algorithms or modify the vendor's model.

The practical consequence is that community banks must conduct meaningful due diligence on vendor AI capabilities, negotiate for transparency and audit rights, monitor vendor AI performance on an ongoing basis, and maintain contingency plans for vendor AI failures. Chapter 9 provides a comprehensive vendor AI governance program, including due diligence questionnaires and model contract provisions.

THE "DEFAULT ON" PROBLEM

One of the most insidious vendor governance challenges involves AI features that are activated by default in software updates. A core processor that introduces an AI-powered fraud detection module in a routine upgrade may not clearly communicate this change, and the bank may not recognize that it is now "using AI" in a regulated function. Community banks must establish procedures for reviewing vendor updates for AI components and assessing their governance implications before activation.

4.5 Data Privacy: GLBA, State Laws, and Emerging Standards

The Gramm–Leach–Bliley Act (GLBA) and its implementing regulations require financial institutions to protect customer information and provide privacy notices describing their information-sharing practices. AI creates additional data privacy considerations that extend beyond GLBA's traditional scope. California, Colorado, Texas, Oregon, and a growing list of states now layer comprehensive privacy statutes—several with automated decision-making and profiling provisions—on top of GLBA's baseline, and the state-enforcement posture described elsewhere in this chapter extends here as well.

AI-SPECIFIC PRIVACY RISKS

Data leakage through generative AI: A deposit operations analyst who uploads a reconciliation spreadsheet to a free AI tool has transmitted customer data to a third-party platform—a failure of the bank's GLBA safeguards obligations in one paste.

Inference and profiling: AI systems can infer sensitive information from non-sensitive data—for example, predicting health conditions from spending patterns. This raises questions about whether such inferences constitute "collection" of sensitive data under privacy laws.

Data retention: AI models trained on customer data retain patterns from that data even after the original records are deleted, creating challenges for data retention and deletion policies.
Cross-border data flows: Cloud-based AI vendors may process data across jurisdictions, potentially subjecting the bank to multiple privacy regimes.

STATE PRIVACY LAWS

A growing number of states have enacted comprehensive privacy laws that may apply to community banks, even those that operate in a single state. These laws often include provisions on automated decision-making, data minimization, and consumer rights that intersect directly with AI governance. The practical step is an annual scope check: have counsel or your compliance officer confirm which state regimes reach the bank, and fold any automated-decision provisions into the Chapter 8 risk assessment for affected systems.

4.6 Emerging Federal and State AI Legislation

No comprehensive federal AI law has been enacted as of this writing, but the bills keep coming—and several developments belong on the bank's radar.

At the federal level, the executive-order landscape has turned over: the 2023 executive order on safe, secure, and trustworthy AI was revoked in January 2025, and subsequent orders have prioritized AI innovation, including a December 2025 order directing a federal review of state AI laws. The federal banking agencies have folded AI into existing guidance and committed to a request for information on generative and agentic AI. Congressional proposals continue to advance—including a 2026 bipartisan draft that would preempt state AI laws for three years—though the timing and scope of any legislation remain uncertain.

At the state level, several jurisdictions have enacted or proposed legislation addressing AI transparency, bias testing, and automated decision-making. Some of these laws apply broadly to businesses, including banks, while others are specific to certain sectors or applications. Appendix I's matrix—kept current in the Toolkit—is the tracking instrument; assign it a named owner rather than a committee.

The practical advice is this: Do not wait for specific legislation to govern AI. Build your governance program on the principles and existing regulatory requirements described in this book, and design it with the flexibility to adapt as new requirements emerge. A bank with a strong foundational governance program can incorporate new regulatory requirements incrementally; a bank starting from scratch will face an urgent and expensive retrofit.

4.7 Examination Preparedness

AI is already an examination topic, and its share of the conversation grows every cycle. Whether the examination is focused on safety and soundness, compliance, IT, or BSA/AML, expect questions about AI use and governance.

WHAT EXAMINERS LOOK FOR

Based on published guidance and examination trends, examiners evaluating AI governance typically assess the following:

- **Awareness:** Does the bank know what AI systems it uses, including vendor-embedded AI?
- **Risk assessment:** Has the bank assessed the risks of its AI systems using a structured methodology?
- **Policies and procedures:** Does the bank have written policies governing AI use, development, and vendor management?
- **Oversight structure:** Is there appropriate management and board oversight of AI risk?
- **Monitoring:** Does the bank actively monitor AI system performance and outcomes?
- **Documentation:** Can the bank produce documentation supporting its governance activities?

THE EXAMINATION CONVERSATION

The banks that fare best in AI-related examinations are those that can tell a coherent governance story. That story begins with awareness ("Here is our AI inventory"), moves through assessment ("Here is how we evaluated the risks"), demonstrates controls ("Here are our policies and monitoring processes"), and shows oversight ("Here is how management and the board stay informed").

A SHIFTING SUPERVISORY POSTURE

The supervisory tone surrounding all of this shifted in late 2025. In its Statement of Supervisory Operating Principles (November 2025, updated April 2026), the Federal Reserve directed examiners to concentrate on a firm's material financial risks and not to devote excessive attention to processes, procedures, and documentation that pose no material risk to safety and soundness. Around the same time, the OCC and FDIC jointly proposed a rule (October 2025) that would, for the first time, define an "unsafe or unsound practice"—broadly, a practice likely to cause material harm to a bank's financial condition—and would tie Matters Requiring Attention (MRAs) more closely to material financial risk or actual violations of law.

For a community bank, the signal is encouraging but easily misread: examiners are being told to weigh AI and model-governance gaps by their actual risk rather than to write up every missing document—but the expectation that you govern AI has not gone away. A gap that creates real fair-lending, financial, or safety-and-soundness exposure is precisely what still draws scrutiny. The approach this book recommends holds up well under the new posture: governance that is proportionate, documented, and defensible reads to an examiner as genuine risk management.

Preparation Checklist

Before your next examination, ensure you can produce: (1) a complete AI/model inventory including vendor-embedded systems, (2) risk assessments for each system proportionate to its risk tier, (3) board-approved AI governance policies, (4) evidence of ongoing monitoring and oversight, (5) documentation of any fair lending or bias testing, and (6) evidence of board and committee engagement with AI governance. The frameworks in the Appendices support each of these elements, and Appendix I extends this list into the full examiner request set, mapped to the chapters that produce each document.

Chapter Summary

This chapter has provided a detailed map of the regulatory landscape as it applies to AI in community banking. The key insights are:

- Existing regulations—model risk management, fair lending, third-party risk, BSA/AML, and data privacy—collectively create comprehensive expectations for AI governance, even in the absence of AI-specific rules.
- Fair lending represents the highest-stakes regulatory domain for AI in banking, requiring proactive testing, monitoring, and documentation.
- Vendor management guidance applies directly to AI acquired through vendor relationships, and banks cannot outsource accountability.
- State AI legislation has already arrived—new state AI laws, state attorney general enforcement actions, and evolving federal priorities all reinforce the need to build flexible governance programs that can absorb new requirements as they come.
- Examination preparedness depends on the ability to tell a coherent governance story supported by documentation.

Looking Ahead

With the regulatory landscape mapped, Chapter 5 turns to the obligations that carry the highest stakes for community banks—AI ethics, bias, and fair lending—translating legal requirements into practical testing, mitigation, and communication. Part III then takes up the work of building the governance program: the organizational architecture of roles, committees, reporting lines, and decision authorities that brings governance to life.

The Exam-Day Reference

The Regulatory Reference Guide in Appendix I maps every requirement to a governance action. The AI Governance Toolkit provides it as an editable matrix you can update as guidance evolves—so your exam prep stays current without rebuilding from scratch.

"An AI system that is accurate but unfair is a liability waiting for an examiner to find it."

— COMMUNITY BANK FAIR LENDING OFFICER —

✦ CHAPTER 5 ✦

AI Ethics, Bias, and Fair Lending

MOVING BEYOND PRINCIPLE STATEMENTS TO OPERATIONAL REALITY

Chapter 3 established the six core governance principles. This chapter puts the ethics-facing ones to work—fairness, transparency, accountability, and privacy—translating principle statements into operational practices. It provides practical methodologies for detecting and mitigating bias, tailoring AI explanations to different audiences, communicating with customers about AI use, and evaluating the community impact of AI adoption.

Community banks have more at stake in this chapter than in any other, because it addresses the qualities that define community banking: fairness, trust, and commitment to community. The community bank that demonstrates responsible AI practices goes beyond compliance—it reinforces the values that differentiate it from larger, impersonal institutions.

5.1 From Principles to Practice

Every financial institution can publish a set of AI ethics principles. Fewer can demonstrate that those principles actually shape decisions. The gap between principle and practice is where governance succeeds or fails.

Principles become real at four moments in a system's life. During risk assessment, the assessor should evaluate the proposed AI system against each principle and document how the system satisfies—or creates tension with—each one. During vendor selection, the due diligence process should include questions specifically derived from the bank's ethical principles. During monitoring, the KPIs tracked should include metrics that measure ethical performance: fairness, explainability, and privacy protection alongside technical accuracy. During incident response, the post-incident review should assess whether the incident reflects a failure of ethical governance as well as a technical one.

The governance committee should periodically—at least annually—review whether the bank's AI practices are consistent with its stated principles. This review should examine specific cases and decisions, looking beyond policies on paper.

5.2 Bias Detection and Mitigation

Bias in AI systems is a measurable phenomenon—one that can be tested for, detected, and addressed. For meaningful bias testing, a structured methodology and a commitment to act on findings matter far more than statistical expertise.

TYPES OF BIAS

- **Historical bias:** The training data reflects past discriminatory practices that the model perpetuates. A lending model trained on data from a period of discriminatory lending will learn to replicate those patterns.
- **Representation bias:** The training data does not adequately represent all customer segments, causing the model to perform poorly for underrepresented groups.

- **Measurement bias:** The data used to measure the outcome the model predicts is itself biased. For example, if historical default data disproportionately reflects certain groups because those groups received less favorable loan terms, the model may conflate the effect of unfavorable terms with group membership.

- **Proxy bias:** The model uses features that correlate with protected characteristics, even though the protected characteristics themselves are not model inputs. ZIP code, for example, often correlates strongly with race and ethnicity.

TESTING METHODOLOGY

The statistical details vary by application; the sequence is the same.

1. Identify the protected classes relevant to the decision—race, ethnicity, sex, age, and other categories protected by applicable law.

2. Analyze the AI system's outcomes across these groups. For lending, compare approval rates, pricing, terms, and other decision factors.

3. Assess whether observed disparities are meaningful, using a widely used first screen: the adverse impact ratio—the protected group's favorable-outcome rate divided by the reference group's—tested against the four-fifths (0.80) threshold and paired with a statistical significance test.

4. Evaluate whether significant disparities can be explained by legitimate risk factors—the business necessity defense.

5. If disparities cannot be explained by legitimate factors, develop and implement remediation strategies.

Note what this test measures: disparity in outcomes. The same arithmetic applies whether a human, a scorecard, or a model made the decisions. What makes a result informative about the model is the population you test.

Run the analysis on the applications the model itself decided, and a flagged result carries a short suspect list—the model and the data feeding it. Run it again on the same period's human-decided applications, and the comparison localizes the cause: a disparity in one stream but not the other tells the investigation where to look.

Where staff can override the model's recommendation, test both views—what the model recommended and what the customer ultimately received. Overrides that consistently favor one group are themselves a fair lending finding.

For community banks with limited statistical resources, third-party fair lending testing services can provide this analysis at reasonable cost. The investment in periodic testing is small compared to the cost of a fair lending finding. Case Study 3 tells the story of a bank that learned this the hard way—and what rebuilding looked like.

Who Performs This Testing

The compliance / fair-lending function owns the testing—scoping it, interpreting the results, and documenting the outcome—and must be independent of the lending business line. The statistical work itself, however—protected-class proxy estimation, regression, significance testing, and the less-discriminatory-alternative search—exceeds the in-house capability of most community banks and is routinely co-sourced to a qualified fair-lending consultant. Fair lending carries real legal exposure, so many banks run any analysis that surfaces a material disparity under the direction of counsel, often under privilege. Keep a non-privileged record of testing cadence, scope, and remediation status—an examiner who cannot see your monitoring will assume there is none. The compliance officer is not expected to compute these results personally—only to own the test and record who performed each part.

A word on the legal ground beneath this test. In April 2026 the CFPB finalized a rule removing the disparate-impact standard from Regulation B, effective July 21, 2026 (Chapter 4 covers the shift in full). That change narrows one federal theory while leaving the bank's actual exposure largely intact. The

> Fair Housing Act's disparate-impact standard still reaches mortgage lending; a growing number of states maintain their own disparate-impact provisions and enforce them through increasingly assertive attorneys general; and disparate outcomes remain a safety-and-soundness and reputational concern regardless of which federal theory applies. The adverse-impact ratio and the four-fifths screen therefore remain the right tools, and periodic testing remains the prudent course. A bank that stopped testing because one federal standard lapsed would be defending against the wrong risk.

MITIGATION STRATEGIES

When testing finds bias, the bank chooses its response by the source and severity of the problem. It can retrain the model on corrected or augmented data. It can remove or adjust the features acting as proxies for protected characteristics. It can build fairness constraints into the model's optimization, or apply post-processing adjustments to the model's outputs. It can put a human reviewer on the borderline cases. And when the bias is severe, it can replace the model altogether.

Mitigation decisions should be documented, including the rationale for the chosen approach and evidence that the mitigation was effective.

> **The Ongoing Nature of Bias Management**
> Customer populations shift, data distributions change, and model behavior drifts—so bias testing recurs. Conduct it at least annually for High- and Critical-tier AI systems, and whenever material changes occur to the model, its data, or the business environment.

5.3 Explainability for Different Audiences

A single AI system may need to be explained to four different audiences, each requiring a different level of detail and framing—the principle Section 3.2 established, applied here audience by audience.

CUSTOMERS

Customers need to understand the basis for decisions that affect them, in plain language. For adverse credit decisions, this means specific, meaningful reasons. "Your application did not meet our criteria" fails that test; "your debt-to-income ratio exceeded our threshold" or "your credit history showed recent late payments" passes it. The explanation should be accurate, understandable, and actionable—the customer should know what they can do to improve their standing.

REGULATORS

Examiners need to understand the model's design, inputs, logic, and performance in sufficient detail to assess its risk. This requires technical documentation, but the documentation should emphasize risk management rather than algorithmic elegance. Examiners want to see that the bank understands the model, has validated it, monitors it, and has controls in place for when it fails.

THE BOARD

The board needs AI decisions translated into business risk language. Instead of model performance metrics, the board needs answers to questions like: Is this system making good decisions? How do we know? What could go wrong? What are we doing about it?

INTERNAL STAFF

Employees who interact with AI outputs—loan officers reviewing AI credit recommendations, compliance analysts reviewing AI-generated alerts, customer service staff supported by AI chatbots—need to understand the system well enough to exercise judgment. They need to know when to trust the AI's output, when to question it, and when to override it. Training and reference materials should be calibrated to this practical need.

5.4 Customer Communication and Transparency

Community bank customers increasingly interact with AI systems, whether they know it or not. The bank's approach to customer communication about AI should be guided by two principles: honesty and proportionality.

WHEN TO DISCLOSE AI USE

At a minimum, disclosure is legally required in specific contexts: adverse action notices must accurately describe the basis for credit denials, regardless of whether a human or algorithm made the decision. A growing number of state AI laws—several effective in 2026—add their own disclosure duties, including telling customers when they are interacting with an automated system rather than a person. Beyond legal requirements, proactive transparency builds the trust that community banks depend on. When customers learn about AI use from a regulator, a news article, or a competitor's marketing rather than from their own bank, trust erodes.

HOW TO COMMUNICATE ABOUT AI

Communication about AI should be clear, honest, and free of jargon. Avoid euphemisms and techno-speak. Saying, "We use technology to help us serve you better," is evasion dressed as transparency. Saying, "We use an AI-assisted system to help evaluate loan applications, and every decision is reviewed by one of our loan officers before it is finalized," is honest, specific, and reassuring. Customers deserve to know that a model is involved—even if they never learn how it works—along with what data it uses, and how they can get a human review if they disagree with the outcome.

5.5 Community Impact Considerations

Community banks exist to serve their communities, and AI adoption should be weighed by its impact on the people and businesses the bank serves, alongside business performance and regulatory compliance.

ACCESS AND INCLUSION

AI-driven services should expand access to banking. The governance program should evaluate whether AI systems might inadvertently exclude certain customer segments—elderly customers who are less digitally engaged, small businesses in underserved areas, customers with limited English proficiency, or individuals with thin credit files who may be penalized by models trained on conventional credit data.

ECONOMIC IMPACT

When AI automates functions previously performed by bank staff, the bank should consider the employment impact on the community. Community banks are often significant employers in their markets, and workforce reductions can have ripple effects. Automation can proceed; the transitions deserve thoughtful management—through retraining, redeployment, and attrition rather than abrupt layoffs where possible.

DIGITAL DIVIDE AWARENESS

AI-powered digital services provide real convenience for digitally connected customers, but they can widen the gap for customers who lack reliable internet access, digital literacy, or modern devices. Community banks should ensure that AI-enhanced digital channels supplement rather than replace human service channels, maintaining the accessibility that defines community banking.

> **The Community Banking Difference**
> Large banks can afford to optimize for efficiency and let community impact be someone else's concern. Community banks cannot. The banker who automates a lending decision will see the affected borrower at the grocery store. The board member who approves an AI strategy will see its impact on the community they live in. This proximity is a powerful natural governance mechanism—one that should be honored as technology advances.

Chapter Summary

This chapter has translated ethical principles into operational practices for community banks:

- Ethical principles become operational through embedding them at specific decision points: risk assessment, vendor selection, monitoring, and incident review.
- Bias detection follows a structured methodology—identify protected classes, analyze outcomes, test significance, evaluate business necessity, and remediate.
- Explainability must be tailored: plain language for customers, risk-focused documentation for regulators, business language for the board, practical guidance for staff.
- Customer communication about AI should be honest, clear, and proportionate—disclosing AI use in chatbots, decisions, and recommendations.
- Community impact assessment ensures AI adoption supports the bank's mission of inclusion, access, and community service.

Looking Ahead

With principles and regulatory obligations grounded in operational practice, Part III turns to building the governance program itself. Chapter 6 designs the organizational structure—the three lines of defense, the governance committee, and the roles and reporting lines that make accountability real. The fair lending discipline built here returns in Section 15.1, where it meets the lending models that need it most.

Make Fairness Measurable

The principles in this chapter become operational through structured review. The AI Governance Toolkit pairs the chapter with three working documents: the Fair-Lending & Bias-Testing Protocol (Section 5.2's methodology as a step-by-step procedure), the AIR Calculator (an Excel workbook that computes the adverse impact ratio, applies the four-fifths screen, and returns a Red/Amber/Green disposition), and the Fair-Lending Test Record (the fill-in evidence form for each test).

PART III

Building the Governance Program

"The best governance structures are invisible to the people they protect and unmistakable to the people they hold accountable."

— COMMUNITY BANK CHIEF RISK OFFICER —

✦ CHAPTER 6 ✦

Designing Your AI Governance Structure

ORGANIZATIONAL ARCHITECTURE FOR OVERSIGHT

Principles provide direction. Regulations provide boundaries. But governance happens through organizational structures: the committees that deliberate, the roles that own accountability, the reporting lines that carry information, and the decision processes that translate principles into action. This chapter designs that organizational architecture.

The goal is a governance structure a community bank can actually run: effective, proportionate, and light enough to survive contact with everyone's day job. We draw on the Three Lines Model, adapt it to AI-specific requirements, and provide ready-to-implement guidance on committee composition, roles, reporting, and integration with existing governance programs.

6.1 The Three Lines Model Applied to AI

The Three Lines Model, maintained by the Institute of Internal Auditors, organizes governance responsibilities. Applied to AI, the three lines operate as follows.

FIRST LINE: BUSINESS OWNERSHIP

The business lines that use AI systems are the first line of defense. This includes lending departments that use AI credit scoring, operations teams that use AI-powered automation, customer service functions that deploy chatbots, and marketing teams that use AI analytics. First-line responsibilities include identifying AI use cases, ensuring day-to-day compliance with AI policies, monitoring AI system performance within their domains, and escalating issues when they arise.

A critical point for community banks: the first line is often where AI adoption begins, sometimes informally. A loan officer who discovers that the new loan origination platform includes an AI recommendation engine is a first-line user, whether or not anyone has formally designated the role. Governance must account for this reality by making first-line AI responsibilities clear, communicated, and supported by training.

SECOND LINE: RISK AND COMPLIANCE OVERSIGHT

The second line is risk management, compliance, and any dedicated AI governance function. Second-line responsibilities include developing AI policies and standards, conducting or overseeing risk assessments, performing independent monitoring and testing (including fair lending and bias testing), advising the first line on governance requirements, and reporting to the governance committee and the board.

In many community banks, the second line for AI governance will be the bank's existing risk and compliance professionals, who add AI oversight to their current responsibilities. This is entirely appropriate, provided these individuals receive adequate AI-specific training and the additional responsibilities are formally documented rather than assumed.

THIRD LINE: INTERNAL AUDIT

Internal audit provides independent assurance that the first and second lines are functioning effectively. For AI governance, this means auditing the AI inventory for completeness, testing whether governance policies are being followed, evaluating the effectiveness of risk assessments and monitoring processes, and assessing whether governance structures are adequately resourced and empowered.

Community banks that outsource internal audit should ensure their audit firm has the competence to evaluate AI governance. Audit-firm capability here is uneven and improving fast; put AI governance scope and methodology on the agenda for your next provider conversation.

> **Practical Reality Check**
>
> In a community bank with 200 employees, the "three lines" may involve the same people wearing different hats. The IT manager may be both a first-line AI user and a second-line policy advisor. This is manageable as long as roles are documented, conflicts of interest are identified, and the governance committee provides independent oversight. Perfection in organizational separation is less important than clarity in accountability.

6.2 The AI Governance Committee: Charter, Composition, and Cadence

The AI governance committee is the operational center of the governance program. In community banks, this may be a new standing committee, a subcommittee of the enterprise risk committee, or an expanded mandate for an existing technology or risk committee (the IT Steering Committee is the most common host). The specific organizational form matters less than the substance: a cross-functional body with the authority, information, and commitment to provide effective AI oversight.

COMMITTEE CHARTER

Every AI governance committee needs a written charter that defines its purpose, authority, membership, and operating procedures. The charter should be approved by the board and reviewed annually. Appendix E provides a charter template. Key elements include:

Purpose: To oversee the bank's use of artificial intelligence, ensure compliance with applicable laws and regulations, manage AI-related risks, and align AI adoption with the bank's strategic objectives and values.

Authority: The committee must have clear decision-making authority, particularly regarding approval of high-risk AI use cases, escalation of AI incidents, and recommendation of policies to the board. A committee that can only advise but not decide will quickly become irrelevant.

Scope: All AI systems used by the bank, including vendor-embedded AI, employee use of generative AI tools, and any AI under development or consideration.

COMMITTEE COMPOSITION

Six chairs need filling—and most banks fill them with people already on the payroll:

IT / Technology: Provides technical expertise on AI systems, infrastructure, and security. Often chairs or co-chairs the committee.

Risk Management: Brings enterprise risk perspective, ensures AI risks are integrated into the bank's overall risk program.

Compliance: Ensures AI governance aligns with regulatory requirements, particularly fair lending, BSA/AML, and consumer protection.

Business Line Representatives: Provide practical perspective on AI use cases, customer impact, and operational needs. Rotate representation across major business lines.

Legal: Advises on contractual, liability, and regulatory interpretation questions. May attend as needed rather than as a standing member.

Senior Management Sponsor: A C-suite executive (often the CIO, CRO, or COO) who ensures the committee has organizational authority and resources.

At banks where one officer covers several functions—IT, information security, compliance, vendor management—five to seven members is usually sufficient. Where those functions are separately staffed, eight to ten. The key is cross-functional representation.

MEETING CADENCE AND STRUCTURE

The committee should meet at least quarterly, with additional meetings as needed for time-sensitive matters such as new AI deployments or AI incidents. Chapter 12 (Section 12.1) provides a detailed model meeting agenda and annual planning calendar. At minimum, each quarterly meeting should cover:

- Review of the AI inventory for changes (new systems, retired systems, vendor updates)
- Status of pending AI use case approvals and risk assessments
- AI risk dashboard review (performance metrics, incidents, monitoring findings)
- Policy and regulatory updates affecting AI governance
- Escalated issues requiring committee decision or guidance

6.3 Roles and Responsibilities Matrix

Clear accountability requires explicit documentation of who is responsible, accountable, consulted, and informed for each governance activity. The following RACI matrix maps key AI governance activities to organizational roles. ***R*** indicates responsibility for performing the activity, ***A*** indicates accountability for the outcome, ***C*** indicates consultation, and ***I*** indicates the role should be informed.

ACTIVITY	BUSINESS LINE	IT / DATA	RISK / COMPL.	INT. AUDIT	GOV. COMM.	BOARD
AI inventory maintenance	C	R	A	I	I	I
Pre-deployment risk assessment	C	R	A	I	A (high)	I
Use case approval (low/mod)	R	C	A	I	I	I
Use case approval (high/crit)	R	C	C	I	A	I
Model validation	I	C	R	C	A	I
Ongoing monitoring	C	R	A	I	I	I
Vendor AI due diligence	C	R	A	I	I	I
AI incident response	C	R	A	I	A (material)	I
Policy development	C	C	R	C	R	A
Board AI risk reporting	I	C	R	I	R	A
AI governance audit	I	C	C	R	I	A

This matrix should be completed to reflect your bank's specific organizational structure, but the principle is constant: every governance activity must have exactly one accountable party. Shared accountability is no accountability. One reading note for the matrix: where a cell shows "A (high)" or "A (material)," accountability transfers from the listed owner to the governance committee—for High- and Critical-tier systems in the first case, for material incidents in the second. The single-accountable-party principle holds within each tier.

> **IT Leader's Note**
> Notice that IT carries "Responsible" designation for many operational governance activities but is not "Accountable" for most of them. This is intentional. IT executes governance activities, but accountability for outcomes rests with risk/compliance (for risk management) and the governance committee or board (for strategic decisions). This structure protects IT from being the sole point of blame when things go wrong and ensures that governance is a shared organizational responsibility.

6.4 Reporting Lines and Escalation Protocols

Information must flow efficiently from the people closest to AI systems to the people responsible for governance decisions. In a community bank, this flow is shorter than in a large institution, but it still requires deliberate design.

ROUTINE REPORTING

Routine AI governance reporting follows a predictable cadence. IT and business line owners provide operational updates to the governance committee quarterly. The governance committee synthesizes these updates into a board-level AI risk report, also quarterly. Risk and compliance functions provide independent monitoring results to both the committee and the board. The cadence gives governance bodies regular, structured information instead of a stream of ad hoc reports.

ESCALATION PROTOCOLS

Not everything can wait for the next quarterly meeting. Escalation protocols define when and how issues should be elevated outside the routine reporting cadence. At minimum, the following events should trigger escalation:

AI system failure: Any AI system failure that affects customers, produces incorrect outputs for more than a defined threshold period, or requires manual override of a critical function.

Bias or discrimination finding: Any testing result or customer complaint suggesting that an AI system is producing discriminatory outcomes.

Data breach or privacy incident: Any unauthorized access to data processed by an AI system or unauthorized transmission of customer data to an external AI platform.

Regulatory finding: Any examination finding, enforcement action, or supervisory communication related to AI.

Vendor material change: Any material change to a vendor's AI system, including model updates, data input changes, or disclosed performance issues.

Each escalation event should have a defined notification chain: immediate notification to the system owner and the risk/compliance function, notification to the governance committee chair within 24 hours, and notification to the board chair or audit committee chair within 48 hours for material events, with the full board briefed at its next meeting. The AI incident response policy (Chapter 7) provides detailed escalation procedures.

6.5 Integrating AI Governance with Existing Programs

AI governance works best bolted onto the programs the bank already runs; five integration points matter most.

ENTERPRISE RISK MANAGEMENT

AI risk belongs in the bank's enterprise risk management program, as a distinct category or as a cross-cutting dimension of the existing ones—operational, compliance, strategic, and reputational. The bank's risk appetite statement should include specific language about AI risk tolerance, and the AI risk dashboard should feed into enterprise-level risk reporting.

INFORMATION SECURITY

AI systems present unique security risks—adversarial attacks, data poisoning, prompt injection—that extend beyond traditional information security concerns. The bank's information security program should be updated to address AI-specific threats, and the CISO or information security officer should be an active participant in AI governance. Chapter 10 addresses AI security threats in detail.

VENDOR MANAGEMENT

Most community bank AI comes through vendors, so AI governance must integrate tightly with the existing vendor management program. The vendor manager adds the AI-specific questions (Appendix C provides them) to the due diligence questionnaire, folds AI risk factors into the vendor risk assessment, and tracks AI performance metrics in ongoing monitoring. Chapter 9 provides a complete vendor AI governance program.

COMPLIANCE MANAGEMENT

AI governance intersects with compliance management across multiple domains: fair lending, BSA/AML, consumer protection, and data privacy. Rather than creating parallel compliance processes for AI, the bank should integrate AI compliance into its existing compliance management system, adding AI-specific testing, monitoring, and reporting where needed.

BUSINESS CONTINUITY PLANNING

If an AI system supports a critical business function, it belongs in the bank's business continuity and disaster recovery plans. For each one, the plan names the people who execute the contingency procedures, the manual fallback they follow, and the recovery time objective they are working against. When the fraud model goes dark, the bank treats it the way it treats any other operational disruption—someone knows what to do, and the plan says who.

> **Integration Principle**
>
> The guiding philosophy is extension. AI governance is most effective when it extends the governance structures the bank already has. Adding an AI section to the vendor management program is more sustainable than creating a separate AI vendor program. Adding AI risk metrics to the enterprise risk dashboard is more effective than building a parallel reporting system. Integration reduces cost, avoids governance fatigue, and ensures that AI risk is managed with the same rigor as every other risk category.

WHAT FOUNDATIONAL GOVERNANCE ACTUALLY COSTS

Proportionality has a number. The estimates below describe the recurring cost of running a Foundational-stage program at a typical community bank once the initial 90-day build is complete. They are planning figures, and they make the central point: this is a part-time discipline layered onto existing roles.

ROLE	TYPICAL COMMITMENT (FOUNDATIONAL STAGE)	WHERE THE HOURS GO
Governance committee members (5–7 people)	3–4 hours per quarter	Quarterly meeting, materials review, approvals
Committee chair	1–2 days per quarter	Agenda, dashboard, board report preparation
IT leader / CISO	2–4 hours per month	Inventory upkeep, monitoring review, vendor follow-up
Compliance / risk officer	2–3 hours per month	Risk assessments, policy maintenance, testing coordination
System owners (first line)	1–2 hours per month per High/Critical system	Performance review, KPI tracking, change-trigger evaluation
Board of directors	30–45 minutes per quarter	Report review and discussion; one annual education session
All staff	1–2 hours per year	Training and attestation

The build itself—the Quick Start's first quarter—is the larger lift; expect it to absorb roughly a year's worth of the table's hours, compressed into ninety days. After that, the table above is the run rate, and it scales with the maturity roadmap in Section 17.4 as the bank's AI footprint grows. The figures assume a footprint of mostly Tier 1–2 systems and a small number of Tier 3; each Tier 4 system adds its own monitoring load on the Section 8.4 schedule, and episodic events—a new High-tier deployment, the annual validation cycle, the annual policy review—arrive as projects on top of the run rate. External engagements are also additional: third-party model validation and fair-lending testing typically run \$15,000–\$50,000 and \$10,000–\$30,000 per engagement (both figures from the Section 10.4 sidebar; Section 5.2 covers the testing itself).

Chapter Summary

This chapter has designed the organizational architecture for AI governance at a community bank:

- The Three Lines Model assigns AI governance responsibilities across business lines (first line), risk and compliance (second line), and internal audit (third line).
- The AI governance committee serves as the operational center, requiring a formal charter, cross-functional membership, and at least quarterly meetings.
- A RACI matrix ensures that every governance activity has clear, singular accountability.
- Reporting lines and escalation protocols ensure efficient information flow with defined triggers for urgent escalation.
- Integration with existing programs—enterprise risk management, information security, vendor management, compliance, and business continuity—ensures sustainability.

Looking Ahead

With the governance structure in place, Chapter 7 turns to the policies that give the structure its teeth: the AI policy suite that defines what is permitted, what is required, and what is prohibited across every dimension of AI use.

Chartering the Committee

The Committee Charter in Appendix E covers purpose, authority, membership, and responsibilities. The AI Governance Toolkit provides it as an editable Word document, ready to complete and present for board approval.

"A policy nobody reads becomes evidence
—proof the bank saw the risk and left it on paper.
A policy that people actually follow is governance."

— BANKING COMPLIANCE CONSULTANT —

✦ CHAPTER 7 ✦

The AI Policy Suite

DRAFTING THE ESSENTIAL POLICIES

Chapter 6 built the organizational structure for AI governance. This chapter fills that structure with substance: the seven essential policies that together define the rules of engagement for AI at your community bank.

Each one addresses a specific governance gap, and an unfilled gap is where findings, losses, and headlines start. They are designed to be adopted as a coordinated suite—each policy references and reinforces the others—while remaining modular enough that a bank can implement them incrementally, beginning with the policies most relevant to its current AI risk profile.

This chapter provides the purpose and scope for each policy, key provisions that must be included, practical drafting guidance, and common pitfalls to avoid. Complete versions for several of these policies are available in the Appendices.

POLICY	PRIMARY PURPOSE	OWNER	WORKED E XAMPLE	WHEN TO ADOPT
AI Governance	Anchor governance structure and define permitted AI use	Risk / Compliance	Appendix A	First 90 days
AI Risk Assessment	Classify and assess AI risk	Risk Management	Appendix B	Quarter 2
Model Development & Validation	Standards for AI model life cycle	IT / Risk	Provisions in §7.3	Quarter 3
AI Vendor & Third-Party	Due diligence and monitoring	Vendor Mgmt / Risk	App. C (questionnaire)	Quarter 2
Data Governance for AI	Data quality, access, and ethics	IT / Data	Provisions in §7.5	Quarter 3
AI Incident Response	Escalation and remediation	Risk / IT	Provisions in §7.6	Quarter 2
Generative AI Acceptable Use	Controls for LLMs and GenAI tools	IT / Compliance	Appendix G	First 90 days

7.1 The AI Governance Policy

The AI Governance Policy is the umbrella document of the suite—and its acceptable-use provisions make it the most broadly applicable. It establishes the ground rules for how every employee, contractor, and vendor may use artificial intelligence in connection with the bank's business, and it anchors the governance structure those rules depend on. Without this policy, AI adoption is ungoverned by default—employees make individual judgments about what is appropriate, with no consistent standard and no organizational accountability.

KEY PROVISIONS

Seven provisions do the policy's heavy lifting:

1. **Scope and applicability:** The policy must clearly state that it applies to all personnel, including temporary staff, contractors, and consultants, and to all forms of AI including generative AI, agentic AI, machine learning, RPA, and AI features embedded in approved software.

2. **Approved AI tools and platforms:** Maintain a list of AI tools approved for use, including any restrictions on how they may be used. Employees should understand that any AI tool not on the approved list requires prior authorization.

3. **Prohibited activities:** Explicitly prohibit high-risk activities such as entering customer personally identifiable information (PII) into unapproved AI tools, using AI to make final credit decisions without human review, relying on AI-generated content in regulatory submissions without verification, and using AI tools that have not been vetted through the vendor management process.

4. **Data handling requirements:** Specify what types of data may and may not be processed by AI tools. At minimum, customer PII, nonpublic personal information, and confidential business information should never be entered into external AI platforms unless specifically authorized.

5. **Disclosure obligations:** Define when and how AI use must be disclosed to customers, particularly in customer-facing applications like chatbots and AI-assisted lending.

6. **Accountability:** State clearly that employees remain responsible for the quality, accuracy, and compliance of their work product, regardless of whether AI tools were used in its creation.

Governance structure and decision rights: Incorporate the governance backbone—the AI Governance Committee's authority, the four risk tiers, and named system ownership—so that the rules and the structure from Chapter 6 are bound in one board-approved document. Appendix A shows the complete twelve-section structure.

DRAFTING GUIDANCE

Write the policy's acceptable-use provisions in plain language that every employee can understand. Use concrete examples of permitted and prohibited behavior—employees comply with examples they recognize from their own desks. Review and update the policy at least annually, and do so more frequently if new AI tools are introduced or new risks emerge.

> **Common Pitfall**
> The most common drafting error is making the policy so restrictive that employees circumvent it rather than comply. If the policy prohibits all use of generative AI but provides no approved alternative, employees will use it anyway—they will simply hide it. Effective policy provides both boundaries and pathways: clear rules about what is prohibited and clear access to approved tools that meet legitimate business needs.

7.2 AI Risk Assessment and Classification Policy

This policy establishes the methodology for assessing and classifying the risk of individual AI systems. It puts the proportionality principle from Chapter 3 to work: higher-risk systems receive more intensive governance, while lower-risk systems receive proportionately lighter oversight.

THE FOUR-TIER CLASSIFICATION SYSTEM

The four tiers—Minimal, Moderate, High, and Critical—are defined by the scoring methodology in Chapter 8, which this policy adopts by reference. The policy itself establishes three things: every AI system receives a tier at intake, a named role assigns and approves the tier, and each tier carries defined governance requirements—from a light annual review at Tier 1 to full validation, enhanced monitoring, and committee approval at Tier 4. A bank debating whether the chatbot is Moderate or High is forcing itself to name exactly what it is worried about—which is the point of the exercise.

CLASSIFICATION CRITERIA

Chapter 8 provides the detailed risk assessment methodology. The classification policy should establish the criteria used for tiering, which typically include customer impact, regulatory sensitivity, data sensitivity, decision autonomy, and model complexity. The policy should also specify who has authority to assign risk tiers, how tier assignments are documented, and how disagreements about tier assignments are resolved.

7.3 AI Model Development and Validation Policy

While most community banks do not develop AI models in-house, this policy is still essential. It establishes standards that apply both to any internal development efforts and to the bank's expectations for vendor-developed models. It also provides the process for model validation, which is a regulatory expectation regardless of where the model was built.

DEVELOPMENT STANDARDS

For any AI model development, including customization of vendor platforms, the policy makes the developing team prove its case before deployment. The team documents the business purpose and intended use. It documents the training and testing data and assesses its quality. It explains the methodology and why that algorithm was chosen, states the performance metrics and acceptance criteria, and records the model's limitations and known weaknesses. Then the appropriate authority approves—before the model goes live.

VALIDATION REQUIREMENTS

In practice, most community banks will outsource—or co-source—the independent validation of their higher-risk models: a qualified external validator performs the quantitative testing, while a designated internal owner scopes the engagement, reviews and challenges the findings, and integrates them into the program. Recognizing that the bank lacks in-house validation capability, and sourcing that capability externally, is sound governance in its own right; what examiners expect is rigorous, independent validation, regardless of who performs it.

Model validation must be performed by individuals or teams independent of the model development and business ownership functions. The policy should define:

- **Scope:** conceptual soundness, data integrity, outcome testing, and backtesting.
- **Frequency:** initial and periodic, calibrated to risk tier.
- **Ad-hoc triggers:** material model changes, performance degradation, or regulatory concerns.
- **Documentation:** validation findings and management responses.

> **Vendor Model Validation**
> A vendor validating its own model is grading its own homework. The vendor's validation tells you the model works in general; your obligation is to know it works for your portfolio, your customers, and your risk profile. That means reviewing the vendor's validation documentation, conducting independent performance testing using the bank's own data, and—for higher-risk models—engaging a third-party validator. Chapter 9 addresses vendor model validation in detail.

7.4 AI Vendor and Third-Party Policy

Given that the majority of community bank AI comes through vendor relationships, this policy is the most operationally important in the suite. It extends the bank's existing vendor management program to address AI-specific risks.

KEY PROVISIONS

Five provisions carry it:

- **AI-specific due diligence requirements:** Define the information the bank must obtain from AI vendors before entering into or renewing a relationship: model documentation, training data descriptions, bias testing results, performance metrics, security controls, and regulatory compliance posture.

- **Contractual requirements:** Specify the AI-related provisions that must be included in vendor contracts: transparency commitments, audit rights, model change notification, data handling obligations, performance guarantees, and termination provisions related to AI governance failures.

- **Ongoing monitoring:** Define how the bank will monitor vendor AI performance after deployment, including periodic performance reviews, alert effectiveness analysis, customer outcome testing, and vendor governance assessment.

- **Material change management:** Require vendors to notify the bank before making material changes to AI models or data inputs, and establish the bank's process for evaluating such changes before they take effect.

- **Concentration risk assessment:** Include provisions for assessing and managing the risk of excessive dependence on a single AI vendor.

Appendix C provides a detailed AI vendor due diligence questionnaire with 25 critical questions organized by category.

7.5 Data Governance for AI

AI inherits its data, flaws included. This policy establishes the data governance standards specific to AI, complementing the bank's broader data governance and information security programs.

CORE REQUIREMENTS

- **Data quality standards:** Define minimum data quality requirements for AI training and input data, including accuracy, completeness, timeliness, and consistency.

- **Data access controls:** Specify who may access data used by AI systems, how access is authorized and revoked, and how access is logged and monitored.

- **Data lineage and provenance:** Require documentation of where data comes from, how it is transformed, and how it flows through AI systems—essential for troubleshooting AI issues and demonstrating compliance.

- **Data retention and deletion:** Address the unique challenges of data retention in the AI context, including the persistence of data patterns in trained models.

- **Prohibited data uses:** Explicitly prohibit the use of data in AI systems in ways that would violate privacy commitments, fair lending obligations, or the bank's own ethical standards.

- **Data minimization:** Require that AI systems use only the data necessary for their intended purpose, avoiding the temptation to collect and process data simply because it is available.

7.6 AI Incident Response Policy

When AI systems fail—and they will—the bank needs a pre-established response policy. Ad hoc responses to AI incidents are slower, less effective, and more likely to compound the damage. This policy defines what constitutes an AI incident, how incidents are reported and escalated, who is responsible for response activities, and how incidents are documented and remediated.

INCIDENT CATEGORIES

The policy should define categories of AI incidents with escalation requirements calibrated to severity:

Category 1—Minor: Performance degradation, non-material output errors, or isolated anomalies. Handled by the system owner with notification to risk/compliance.

Category 2—Significant: Customer-affecting errors, material performance failures, or data integrity issues. Requires governance committee notification and formal incident report.

Category 3—Critical: Discriminatory outcomes, data breaches, regulatory violations, or widespread system failures. Requires immediate escalation to senior management and the board, regulatory notification where required—including the federal Computer-Security Incident Notification Rule's 36-hour notice to the bank's primary regulator when an incident rises to that level—and activation of the bank's crisis management procedures.

RESPONSE PROCEDURES

For each category, the policy answers the questions an incident raises while it is still happening. Who has the authority to shut the system down, and what containment steps come first. How the bank investigates and runs the root cause to ground. What must be remediated, and on what timeline. Who tells whom, inside the bank and outside it. What gets documented, so the record stands up in the next examination. And what the post-incident review changes, so the controls come out stronger.

7.7 Generative AI Acceptable Use Policy

Generative AI—large language models, image generators, and related tools—presents a distinct set of governance challenges that merit a dedicated policy. While the AI Governance Policy's acceptable-use provisions set the broad expectations, generative AI's unique characteristics require additional specificity.

WHY A SEPARATE POLICY?

Generative AI differs from traditional AI in several ways that matter for governance. These tools are widely accessible to any employee with internet access, creating a large attack surface for data leakage. They produce plausible-sounding but potentially inaccurate outputs, creating reliability risks in professional contexts. They can generate content that inadvertently violates intellectual property rights. And their underlying models are updated frequently by their providers, making performance unpredictable over time.

KEY PROVISIONS

- **Approved platforms:** List the specific generative AI tools approved for bank use and any restrictions on their use (for example, an enterprise version with data protection features versus a free consumer version).

- **Data input restrictions:** Strictly prohibit entering customer PII, nonpublic personal information, confidential business data, examination materials, SARs and SAR-supporting investigation materials—whose confidentiality carries criminal exposure under the Bank Secrecy Act—or any other sensitive information into generative AI tools unless specifically authorized.

- **Output verification:** Require that all generative AI output used in bank business be reviewed and verified by a qualified human before use. No generative AI output should be presented to customers, regulators, or the board without human review.

- **Attribution and disclosure:** Define when and how the bank discloses that generative AI was used in creating content, and require that AI-generated content not be represented as original human work in contexts where that distinction matters.

- **Training and awareness:** Require all employees to complete generative AI training before using approved tools, covering both the capabilities and the limitations of these systems.

Board Perspective

The generative AI policy is the one employees test every day, mostly without realizing it—every paste into a chatbot is a policy decision made at a desk. Expect management to report three numbers: adoption, training completion, and violations. A violations column that reads zero for four straight quarters usually means the measuring stopped. This is the policy where culture and training matter as much as the text.

7.8 Anticipating and Addressing Policy Resistance

Every AI policy in this chapter will face pushback. That pushback is a sign the policy is reaching the people who actually use AI in their daily work. The governance committee member who cannot anticipate and address resistance will find policies adopted on paper but ignored in practice. This section prepares you for the five most common objections and provides responses grounded in both pragmatism and regulatory reality.

"WHY CAN'T I JUST USE CHATGPT ON MY PHONE? IT'S MY PERSONAL DEVICE."

This is the single most common objection, and it reveals a genuine misunderstanding. The policy does not regulate what employees do on their personal devices for personal purposes. It regulates what employees do with bank data and bank business, regardless of the device. When a loan officer pastes a customer's tax return into a personal ChatGPT account on their phone, the bank's data has left the bank's control. Whatever the device, the data belongs to the bank and its customers.

The effective response combines empathy with clarity: "We understand these tools are helpful—that's exactly why we've provided an approved enterprise platform that gives you the same capabilities with the data protections our customers expect and our regulators require. Use the approved tool, and you can get the benefits without the risk."

"THIS SLOWS US DOWN. WE WERE MORE PRODUCTIVE BEFORE THE POLICY."

This objection is sometimes accurate in the short term. Adding a human review step to AI-generated customer communications does take more time than sending them unreviewed. The response should acknowledge this honestly while reframing the trade-off: the policy exists to prevent the far more costly slowdowns that come from regulatory findings, customer complaints, data breaches, and reputational damage. A fair lending violation discovered during an examination will consume hundreds of hours of management time and potentially millions in remediation. The five minutes spent reviewing an AI-generated credit memo is inexpensive insurance.

Where possible, address the productivity concern directly by investing in the approved enterprise tools, streamlining approval workflows, and eliminating unnecessary steps that do not reduce risk. The goal is governance that does its job at the lowest cost in productivity.

"OUR VENDOR MANAGES THE AI. WHY DO WE NEED OUR OWN GOVERNANCE?"

This objection reflects one of the most dangerous misconceptions in community banking. Regulators have been unambiguous: a bank cannot outsource responsibility by outsourcing a function. When a vendor's AI model produces a discriminatory lending outcome, the examiner cites the bank. When a vendor's AI chatbot gives a customer incorrect financial advice, the customer sues the bank.

The response is direct: "We trust our vendors, and we have strong vendor relationships. But trust is not governance. Governance means we understand what the vendor's AI does, we've assessed the risks, we monitor performance, and we have contractual protections. Our examiner will not accept 'the vendor handles it' as a governance response, and neither should we."

"WE'RE TOO SMALL FOR ALL THIS. THIS IS FOR BIG BANKS."

This objection confuses the scale of governance with the need for governance. The book's entire approach is built on proportionality—a community bank needs governance proportionate to its AI use—far lighter than JPMorgan Chase's, and still real. A bank with three AI systems needs simpler governance than a bank with thirty, but it still needs an inventory, risk assessments, policies, and oversight.

The most effective counter to this objection is the examination reality: examiners are asking community banks about AI governance today. The bank that can produce an inventory, a few policies, and evidence of committee oversight is far better positioned than the bank that says, "We're too small for that." The Quick Start guide at the beginning of this book proves that foundational governance can be built in 90 days with existing staff.

"I'VE BEEN DOING THIS FOR TWENTY YEARS. I DON'T NEED A COMPUTER TO TELL ME WHO'S CREDITWORTHY."

This objection comes from experienced bankers who feel that AI undermines their professional judgment—and it deserves respect, because that judgment is genuinely valuable. The response should validate the concern while reframing the relationship between human judgment and AI: "Your experience is exactly why you're essential to this process. The AI model provides one data point—a score based on patterns in historical data. Your judgment provides the context that no model can capture: the character of the borrower, the dynamics of their business, the local economic conditions. The governance program asks you to consider the AI's input alongside your own judgment, document your reasoning, and override the model when your judgment tells you the model is wrong. That keeps your expertise as the deciding voice—and gives it one more tool."

Governance Committee Preparation

Before rolling out any new AI policy, the governance committee should anticipate the three most likely objections from the affected employee group; prepare clear, empathetic responses; brief managers who will handle questions from their teams; and create a brief FAQ document that accompanies the policy. Resistance is manageable when it is anticipated. It becomes corrosive when it catches the organization off guard.

Chapter Summary

This chapter has outlined the seven essential policies in the AI governance suite:

- The AI Governance Policy anchors the governance structure and establishes ground rules for all AI activity across the organization.
- The Risk Assessment and Classification Policy operationalizes proportionality through a four-tier system.
- The Model Development and Validation Policy sets standards for the AI model life cycle.
- The Vendor and Third-Party Policy extends vendor management to address AI-specific risks.
- The Data Governance Policy ensures data quality, access control, and ethical data use.
- The Incident Response Policy pre-establishes response procedures for AI failures.
- The Generative AI Acceptable Use Policy addresses the unique risks of LLMs and related tools.

Looking Ahead

Chapter 8 takes the risk assessment policy from this chapter and builds it into a complete, step-by-step methodology—from discovering every AI system in your organization to conducting pre-deployment assessments and establishing ongoing monitoring.

The Policy Suite, Drafted

Every provision the seven policies need is in this chapter. Appendices A and G show two of them adopted end to end, Appendix B shows what the risk-assessment policy produces—a completed assessment—and Appendix C provides the companion vendor due diligence questionnaire; Sections 7.3, 7.5, and 7.6 carry the full provisions for the rest. The AI Governance Toolkit packages the suite as editable templates formatted for board presentation and regulatory review—five standalone policies plus the working instruments the other two drive—it saves the drafting; the thinking is already in this chapter. See the Toolkit page at the back of this book.

“Our first AI inventory came back with twenty-three systems. Going in, we would have told you four.”

— COMMUNITY BANK IT DIRECTOR —

✦ CHAPTER 8 ✦

AI Risk Assessment: The Four Tiers

FROM INVENTORY TO ONGOING MONITORING

Twenty-three systems, when the bank would have guessed four—the epigraph's arithmetic is typical, and it is the reason this chapter exists. What follows is a step-by-step methodology for discovering every AI system in your organization, classifying each by risk, assessing the ones that matter before they go live, and monitoring them in proportion to their tier. It turns the risk-assessment policy of Chapter 7 into work a community bank can schedule.

This methodology is designed for community banks. It satisfies regulatory expectations with the staff a community bank actually has. Every step includes concrete guidance, and Appendices B and F show what the supporting documentation looks like (editable versions are in the Toolkit).

A Note on Sourcing Your Answers

Many of the fields in these assessments—a model's methodology, the data it was trained on, its assumptions and known limitations—are not things a community bank will know from memory, and you are not expected to. For the vendor and embedded systems that make up most community-bank AI, the answers come from the vendor's documentation, your vendor due-diligence questionnaire, the contract, and your own IT and business owners. Complete what you can, request the rest, and where an answer is unavailable, record it as a documented gap or limitation rather than leaving it blank or guessing. An honest "not provided by the vendor" is defensible to an examiner. A fabricated answer or an empty field invites a finding. A vendor's refusal to provide material information is itself a finding to track.

8.1 The AI Inventory: Finding Every AI System

The AI inventory is the foundation of the entire governance program. Without a complete inventory, risk assessments are incomplete, monitoring is spotty, and examiners will rightfully question the governance program's credibility. Building the inventory is the hardest step in this chapter for one reason: much of the bank's AI is invisible.

DISCOVERY TECHNIQUES

Five techniques, used together, find what no single one will:

1. **Vendor review:** Examine every vendor relationship for AI components. Request AI disclosure from each vendor: "Does your product or platform use artificial intelligence, machine learning, or algorithmic decision-making in any capacity?" Core processors, digital banking platforms, loan origination systems, CRM tools, and compliance monitoring software increasingly embed AI features.

2. **Technology audit:** Review the bank's technology stack for AI-related capabilities: machine learning libraries, API connections to AI services, cloud computing resources used for model training, and automated decision-making workflows.

3. **Employee survey:** Ask employees directly: "Are you using any AI tools in your work, including ChatGPT, Copilot, or similar products?" Make clear that the survey is a count, with no consequences attached. Employees who fear punishment will not disclose shadow AI use.

4. **Process mapping:** Walk through key business processes (lending, account opening, fraud detection, customer service, compliance reporting) and identify where automated decisions or recommendations occur. Any automated decision point is a potential AI system.

5. **Software license review:** Examine software licenses and subscriptions for AI-related tools or features that may have been purchased but not yet inventoried.

INVENTORY DATA ELEMENTS

For each AI system identified, the inventory should capture the system name and description, the vendor (if applicable), the business function and department, the type of AI (ML, RPA, generative AI, agentic AI, etc.), the data inputs, the outputs and their use, the designated system owner, the risk tier assignment, the date of last assessment, and the current status (active, in development, retired). Appendix F provides a spreadsheet-ready inventory format.

The "Default On" Discovery

During the initial inventory, many banks discover AI features that were activated without their knowledge through vendor software updates. One community bank discovered that its core processor had enabled an AI-driven overdraft prediction model three software releases ago. The model was making recommendations to tellers about which customers to offer overdraft protection—a customer-facing AI application operating without any governance whatsoever. Three releases is roughly eighteen months: the model had been advising tellers for a year and a half before anyone in governance knew its name.

8.2 Risk Classification Criteria

Once the inventory is complete, each AI system must be classified by risk. Here the four-tier system from Chapter 7 becomes a scoring method.

SCORING DIMENSIONS

Each AI system is scored across five dimensions on a 1–4 scale. The overall risk tier is determined by the highest individual dimension score: any dimension scoring 4 places the system in Tier 4 (Critical), regardless of other scores. This approach ensures that a single high-risk characteristic is not averaged away.

DIMENSION	MINIMAL (1)	MODERATE (2)	HIGH (3)	CRITICAL (4)
Customer Impact	None / internal only	Indirect influence on customer experience	Directly affects individual customers	Automated decisions with material financial impact
Regulatory Sensitivity	No regulatory nexus	Touches regulated domain, low scrutiny	Fair lending, BSA, or consumer protection	Core to regulatory compliance function

DIMENSION	MINIMAL (1)	MODERATE (2)	HIGH (3)	CRITICAL (4)
Data Sensitivity	Public / non-sensitive	Internal business data	Customer PII, financial records, credit-bureau data	Protected-class, health, or biometric data
Decision Autonomy	Informational only	Recommends; human decides	Decides with human review	Autonomous, limited oversight
Model Complexity	Rules-based / transparent	Simple statistical model	ML model, interpretable	Deep learning / opaque

SCORING AND RATIONALE

Document each system's scores and the reasoning behind them in this format; Appendix B shows the template completed end to end.

RISK DIMENSION	SCORE (1–4)	RATIONALE / NOTES
Customer Impact		
Regulatory Sensitivity		
Data Sensitivity		
Decision Autonomy		
Model Complexity		
Overall Risk Tier		

TIER ASSIGNMENT RULES

Tier 4 (Critical): Any dimension scores 4, OR the system renders consequential customer decisions in credit or BSA/AML with limited human review.
Tier 3 (High): Maximum dimension score is 3, OR the system directly affects customer credit decisions or BSA/AML compliance. Credit and BSA/AML systems enter at Tier 3 as a floor; when such a system acts with limited human review, Decision Autonomy scores a 4 and lifts the classification to Tier 4.
Tier 2 (Moderate): Maximum dimension score is 2, AND the system does not make direct customer decisions.
Tier 1 (Minimal): All dimensions score 1, internal use only, easily reversible.

The governance committee should review and ratify all Tier 3 and Tier 4 classifications. Tier 1 and Tier 2 classifications may be assigned by risk/compliance with periodic committee review. Disagreements about tier assignments should be escalated to the committee, with the higher tier applied pending resolution.

One Method, Two Forms

The Toolkit's AI Inventory & Classification Register operationalizes these five dimensions as two composite scores—inherent risk (model complexity, decision autonomy, data sensitivity) and materiality (customer impact, regulatory exposure)—using the same 1–4 scale, the same highest-score rule, and an automatic Critical lock for BSA/AML systems. Same methodology, fewer columns: the two-axis form speaks the materiality language of SR 26-2 that examiners now use. Where a BSA/AML system only flags activity for human review, a bank may document a High classification in place of the Critical default—the split reflected in the worked example of Appendix F (FraudShield at Critical, the alert engine at High).

8.3 Pre-Deployment Risk Assessment Process

Before any AI system goes live—whether newly acquired, newly developed, or newly discovered in the inventory—it must undergo a risk assessment proportionate to its tier. The assessment process has six steps.

1. **Business Case Review**
   Document the business purpose, expected benefits, and intended users of the AI system. This ensures that the bank has a clear rationale for deploying AI and a baseline against which to measure performance.

2. **Data Assessment**
   Evaluate the data the AI system will use: its source, quality, relevance, and any potential for bias. For vendor systems, request documentation of the training data and any known data quality issues. Assess whether the data is representative of the bank's specific customer base and market.

3. **Model Evaluation**
   For Tier 2 and above, evaluate the AI model's methodology, performance metrics, and limitations. For vendor models, this may rely on vendor documentation supplemented by the bank's own testing. For Tier 3 and Tier 4 systems, engage an independent validator.

4. **Bias and Fairness Testing**
   For any AI system that affects customer outcomes (typically Tier 2 and above), conduct bias testing appropriate to the application. For lending models, this means disparate impact analysis. For customer-facing AI, this means testing for differential treatment across demographic groups. Chapter 5 provides detailed bias testing guidance.

5. **Security and Privacy Review**
   Assess the AI system's security architecture, data handling practices, access controls, and compliance with privacy obligations. For cloud-based or vendor-hosted AI, evaluate the security of the hosting environment and the data transmission methods.

6. **Approval and Documentation**
   Document the assessment findings in a standardized format (Appendix B provides the template), assign the risk tier, and route for approval. Tier 1 and Tier 2 systems may be approved by risk/compliance management. Tier 3 systems require governance committee approval. Tier 4 systems require governance committee approval with board notification.

> **Practical Guidance for Small Banks**
> For banks with limited staff, the pre-deployment assessment does not need to be a separate project for each AI system. Batch similar systems together. If you are evaluating three vendor AI features from the same core processor, conduct a single comprehensive assessment that covers all three, with supplemental analysis for features that differ in risk profile. One assessment covering three systems is proportionality doing its job.

8.4 Ongoing Monitoring and Performance Measurement

AI systems evolve, data distributions shift, and the risk environment changes. Assessment has to keep up. Ongoing monitoring ensures that governance remains current and that emerging risks are detected before they cause harm.

MONITORING REQUIREMENTS BY TIER

The monitoring approach calibrates activity frequency to risk tier:

ACTIVITY	TIER 1 (MINIMAL)	TIER 2 (MODERATE)	TIER 3 (HIGH)	TIER 4 (CRITICAL)
Performance review	Annual	Semi-annual	Quarterly	Monthly
Bias / fair lending test	N/A	Annual (if applicable)	Semi-annual	Quarterly rate monitoring; pooled statistical test semi-annual
Independent validation	N/A	As needed	Risk-triggered	Risk-triggered; at least annual currency review
Drift detection	N/A	Annual	Quarterly	Continuous
Governance committee report	Annual summary	Semi-annual	Quarterly	Quarterly + ad hoc
Board reporting	Part of annual summary	Semi-annual summary	Quarterly	Quarterly + ad hoc

KEY PERFORMANCE INDICATORS

A model never announces its own decline; the KPIs are how the bank notices. While specific KPIs vary by application, common categories include accuracy and error rates, false positive and false negative rates (particularly for fraud and BSA/AML systems), customer outcome metrics (approval rates, denial rates, complaint rates), model stability measures (score distribution shifts, feature importance changes), and operational metrics (uptime, response time, exception volume).

DRIFT DETECTION

Model drift occurs when an AI system's performance degrades over time because the data it processes changes relative to the data it was trained on. Drift detection is particularly important for Tier 3 and Tier 4 systems. Common approaches include monitoring score distributions for shifts, comparing predicted outcomes to actual outcomes over time, tracking feature importance and alerting on significant changes, and conducting periodic backtesting using recent data. When drift is detected, the governance program should trigger a reassessment of the AI system, potentially including revalidation and retraining.

8.5 Change Management for AI Systems

AI systems change more frequently than traditional software. Vendors release model updates, retrain models on new data, modify feature sets, and adjust thresholds. Internally, data inputs evolve as the bank's customer base and product mix change. Each of these changes can alter the AI system's risk profile.

CHANGE TRIGGERS

The change management policy should define events that trigger reassessment:

- **Vendor model updates:** Any change to the model's algorithm, training data, feature set, or threshold parameters.
- **Data input changes:** New data sources, significant changes in data quality or completeness, or changes in the population the model serves.
- **Performance degradation:** Monitoring results that indicate declining accuracy, increasing error rates, or emerging bias.

- **Regulatory changes:** New regulatory guidance or examination findings that affect the AI system's compliance posture.

- **Business changes:** New products, new markets, or significant changes in the bank's risk profile that may affect AI system appropriateness.

CHANGE ASSESSMENT PROCESS

When a change trigger occurs, the system owner should evaluate whether the change is material. Material changes require the same assessment process used for new deployments (Section 8.3), scaled to the change's scope. Non-material changes require documentation and notification to risk/compliance but may not require full reassessment. The governance committee should review the criteria for distinguishing material from non-material changes and approve the assessment outcomes for Tier 3 and Tier 4 systems.

Chapter Summary

This chapter has provided a complete, practical risk assessment methodology:

- The AI inventory process uses multiple discovery techniques to find every AI system, including vendor-embedded and shadow AI.
- A five-dimension scoring method classifies each system into one of four risk tiers, with the highest single dimension determining the tier.
- A six-step pre-deployment assessment process evaluates business case, data, model quality, bias, security, and documentation.
- Ongoing monitoring is calibrated to risk tier, with defined KPIs, drift detection, and periodic reassessment.
- Change management procedures ensure that updates to AI systems trigger appropriate reassessment.

Looking Ahead

Chapter 9 applies these governance principles to the specific context of vendor relationships—the primary channel through which community banks acquire and use AI.

Your First Assessment

Appendix B shows a completed assessment of a Critical-tier credit decisioning system. The AI Governance Toolkit provides the blank template as an editable Word document, in the same format as Appendix B, ready for your first real assessment.

PART IV

Managing the Risks

"Accountability is the one thing an outsourcing contract cannot transfer."

— PARAPHRASE OF OCC GUIDANCE ON THIRD-PARTY RELATIONSHIPS —

✦ CHAPTER 9 ✦

Governing the AI You Buy

MANAGING AI YOU DON'T BUILD BUT STILL OWN THE RISK FOR

Most of the AI a community bank relies on, it does not build. AI features arrive embedded in core processing platforms, machine learning models come packaged in lending software, fraud-detection algorithms run inside payment systems, and generative-AI tools are offered by the technology providers you already work with. This chapter addresses the governance challenges that arise when the AI you depend on is built, maintained, and updated by someone else—and it maps them onto the third-party risk-management life cycle your regulators already expect you to follow.

The fundamental challenge is this: the bank bears regulatory and reputational responsibility for the AI's outcomes, but the bank did not build the AI, may not fully understand it, and may have limited ability to inspect or modify it. The bank bridges that gap with four tools it already understands: due diligence before the contract, protections inside it, monitoring after it, and a contingency plan for the day the relationship ends. Together they let the bank meet its governance obligations even when the AI is someone else's intellectual property.

Vendor AI governance is your existing third-party risk-management program, extended to a new kind of risk. The June 2023 Interagency Guidance on Third-Party Relationships—issued as OCC Bulletin 2023-17, Federal Reserve SR 23-4, and FDIC FIL-29-2023—sets the expectation, and the 2024 companion guide written specifically for institutions with $10 billion or less in assets walks through it stage by stage. That guidance frames third-party risk as a life cycle, and AI fits inside each stage:

- **Planning:** decide whether AI is the right tool for the need, and bring the use case into your inventory before a contract is signed.
- **Due diligence and selection:** the five AI-specific domains in Section 9.1.
- **Contract negotiation:** the provisions in Section 9.3.
- **Ongoing monitoring:** the activities in Section 9.4.
- **Termination:** the ability to exit—data return and deletion, model-output portability, and a transition period that does not strand a critical function.

The machinery already exists in your vendor program. The work of this chapter is adding the AI-specific questions, clauses, and controls at each stage.

9.1 Vendor AI Due Diligence

Standard vendor diligence still matters—financial stability, operational resilience, information security. AI adds dimensions of its own, and they need their own questions.

PRE-ENGAGEMENT ASSESSMENT

Before entering into a vendor relationship that involves AI, or before activating AI features in an existing vendor platform, the bank should assess the vendor across five AI-specific domains:

1. **Model governance maturity:** Does the vendor have its own model risk management program? What is the vendor's approach to model development, testing, validation, and monitoring? A vendor that cannot articulate its own governance practices is unlikely to support your governance needs.

2. **Transparency and documentation:** Can the vendor provide sufficient documentation of the AI system's design, training data, methodology, performance metrics, and limitations to support the bank's risk assessment and regulatory compliance?

3. **Bias testing and fair lending:** Has the vendor tested the AI system for bias and disparate impact? Can the vendor provide testing results? Is the vendor willing to support the bank's own bias testing using bank-specific data?

4. **Data handling:** How does the vendor collect, store, process, and protect data used by the AI system? Where is data processed geographically? Does the vendor use bank data to train models for other clients?

5. **Change management:** How does the vendor manage AI model changes? Will the bank receive advance notification of material model updates? Does the bank have the ability to test changes before they take effect?

The Transparency Negotiation

Many vendors will initially resist detailed AI transparency requests, citing proprietary protection. The bank's response should be firm: "Keep your source code. What we need is the documentation our regulators require us to have." Most vendors, when they understand the regulatory basis for the request, will provide meaningful documentation. Those that cannot or will not should be evaluated with heightened skepticism.

9.2 Key Questions for Vendor Selection

Appendix C contains the complete 25-question due diligence instrument, organized into five categories. The categories matter more than the individual questions—each exists to force a specific kind of disclosure—so this section explains what each category is designed to surface, with an exemplar question. Send the full questionnaire; read the answers against the purpose described here.

Model Transparency (Appendix C, Questions 1–5)

1. Transparency questions force the vendor to document what the model does, what data trained it, and what its known limitations are. This is the foundation for every later obligation, because a bank cannot explain to a regulator what a vendor will not explain to the bank. From the questionnaire (Question 3): "Can you provide documentation explaining the model's decision logic at a level sufficient for our regulatory obligations?"

Bias and Fairness (Appendix C, Questions 6–10)

2. Fairness questions establish whether the vendor has tested for disparate impact, how it remediates findings, and whether testing recurs as the model changes. Resistance to sharing methodology and results is itself a disclosure. From the questionnaire (Question 6): "Has the model been tested for disparate impact on protected classes? What methodology was used and what were the results?"

Data Governance (Appendix C, Questions 11-15)

3. Data questions locate the bank's data and pin down its permitted uses, including the one that surprises banks most: whether customer data trains models offered to other clients. From the questionnaire (Question 12): "Is our customer data used to train or improve models that serve other clients?"

Operations and Resilience (Appendix C, Questions 16-20)

4. Resilience questions test whether the AI is operationally serious—validation practices, performance monitoring, update discipline, incident response, and continuity provisions. From the questionnaire (Question 18): "How do you handle model updates? Will we be notified before material changes take effect?"

Compliance and Governance (Appendix C, Questions 21-25)

5. Governance questions reveal whether the vendor governs itself the way the bank must—a formal program, examination support, audit rights, and community-bank references. Paired with Question 18's change-notification commitment, these are the questions Case Study 2 puts a price on. From the questionnaire (Question 23): "Are you willing to grant us audit rights specific to your AI systems?"

9.3 Contractual Provisions for AI

Due diligence reveals the vendor's current practices. Contracts lock in the protections the bank requires. AI-related contractual provisions should be integrated into the bank's standard vendor contracts, with additional specificity for high-risk AI relationships.

EIGHT ESSENTIAL CONTRACTUAL CLAUSES

1. **Transparency commitment:** The vendor agrees to provide and maintain documentation of methodology, data inputs, performance metrics, and limitations sufficient for the bank's regulatory compliance.

2. **Model change notification:** Written notice at least 30 days before implementing material changes to AI models, with material changes defined contractually.

3. **Audit rights:** The bank retains the right to audit vendor AI governance, testing, and performance either directly or through designated third parties.

4. **Fair lending cooperation:** The vendor agrees to cooperate with the bank's fair lending testing, including providing necessary data and documentation.

5. **Data ownership and handling:** Clear bank data ownership, restrictions on vendor use for other clients' model training, and data deletion upon termination.

6. **Performance guarantees:** Measurable performance standards with remedies including termination rights if thresholds are breached.

7. **Indemnification:** Vendor indemnifies the bank for losses from AI system failures, biased outcomes, or data breaches attributable to the vendor.

8. **Regulatory cooperation:** Vendor agrees to cooperate with the bank's regulators during examinations.

When the Vendor Says No

No bank wins every clause. Rank them. The must-have provisions are model-change notification, audit rights (or their substitute), data ownership and deletion, and fair-lending cooperation; the rest strengthen the bank's position but can be traded. When a vendor refuses a must-have, negotiate to a fallback rather than abandoning the protection:

- **Audit rights refused:** require a current SOC 2 Type II report plus the contractual right to the testing results, and rely on the federal service-provider examination for large cores.
- **30-day model-change notice refused:** require delivery of release notes and a defined window to test and object before a change takes effect.
- **Fourth-party disclosure refused:** require an annual subprocessor attestation.

One provision deserves its own line, because it addresses the way AI most often enters a bank unnoticed: an AI-activation disclosure clause, under which the vendor agrees to notify the bank before enabling any new AI feature in a product the bank already uses. Pair it with a standing review of vendor release notes and a periodic re-inventory, so a capability the bank never switched on does not quietly become a capability the bank cannot account for.

Negotiating Leverage

Community banks often feel they lack negotiating leverage with large vendors. Several strategies improve outcomes: engage in discussions before contract renewal deadlines; participate in industry groups that negotiate collectively; reference specific regulatory guidance to establish that these requests are grounded in regulatory requirements; and remember that walking away is always an option.

9.4 Ongoing Vendor AI Monitoring

Due diligence and contracts are point-in-time protections; monitoring is the rest of the relationship.

MONITORING ACTIVITIES

- **Performance tracking:** Monitor AI system performance using KPIs from the risk assessment. Compare actual performance to contractual commitments.
- **Outcome analysis:** Analyze outcomes produced by vendor AI systems across demographic groups for lending models, alert quality for fraud detection, and complaints for customer-facing AI.
- **Vendor governance review:** Annually reassess the vendor's AI governance maturity and request updated documentation of testing results and incidents.
- **Change assessment:** Evaluate vendor-notified model changes for risk impact before accepting them, with independent testing for material changes to high-risk systems.
- **Regulatory alignment:** Monitor whether vendor practices remain aligned with evolving regulatory expectations as new guidance is issued.

VENDOR SCORECARD

Develop a vendor AI scorecard that consolidates monitoring results into a single assessment, tracking performance reliability, transparency quality, responsiveness, change management discipline, and regulatory alignment. The scorecard provides a structured basis for annual vendor reviews and contract renewal decisions.

9.5 Concentration Risk and Vendor Dependency

Community banks should assess the strategic risk of excessive dependence on a single AI vendor. Concentration risk shows up as single points of failure across multiple business lines, reduced negotiating leverage, and strategic lock-in that makes switching prohibitively complex.

MITIGATION STRATEGIES

While complete diversification may not be practical, several strategies mitigate concentration risk. Maintain manual fallback procedures for every critical AI function. Include strong termination provisions ensuring data portability and reasonable transition periods. Periodically evaluate alternative vendors so the bank always knows what switching would cost. Ensure that the bank's own data is maintained in a format that can be migrated to a new platform if necessary.

> **Board Perspective**
> Vendor AI concentration risk deserves periodic board attention during strategic planning. The board should understand which vendors supply AI for critical functions, what fallback options exist if a key relationship is disrupted, and whether contract terms provide adequate protection. Concentration is the risk that accumulates while every individual decision looks sensible; the board is the only seat with the full view.

9.6 Fourth-Party Risk and Foundation-Model Concentration

Section 9.5 addressed the risk of leaning too heavily on a single vendor. There is a deeper version of that risk, and most banks never see it: the vendors themselves depend on a very small number of foundation-model providers. The lending platform, the fraud system, and the productivity assistant a bank treats as three separate vendors may all run on the same underlying model from one of a handful of providers. Apparent diversity at the vendor layer can mask a single shared dependency at the model layer—one the bank never chose and cannot see.

Researchers have a name for the systemic version of this problem—"algorithmic coupling": when many institutions run the same model architectures, trained on similar data and prone to the same failure modes, their decisions become correlated, so a single model flaw or data-contamination event can affect every bank using it at once. The concern has already reached the top of the system: in April 2026, the Treasury Secretary and the Federal Reserve Chair convened the heads of the largest U.S. banks over the cyber-risk implications of a single new AI model, a reminder that the model layer is a shared dependency no individual bank controls.

For a community bank, the practical response is fourth-party visibility. Ask each AI vendor which foundation-model providers and subprocessors it relies on, require disclosure of those dependencies and notice when they change, and flow your own AI obligations—testing cooperation, incident notification, change notice—down to the model layer through the contract. Know your fallback if an underlying model is restricted, withdrawn, or compromised. The Financial Services AI Risk Management Framework treats subprocessor disclosure and model-change notification as baseline control objectives; your vendor questionnaire should too.

9.7 Embedded AI and the Low-Leverage Reality

Not every vendor relationship gives the bank room to negotiate. It helps to separate two situations the rest of this chapter has treated as one.

A procured model is one the bank selects and could, in principle, test, document, and replace—a lending or fraud model weighed against alternatives. Here the due-diligence questions and contract provisions in this chapter apply with full force.

An embedded AI feature is different: an AI capability switched on inside a core processing, digital banking, or productivity platform the bank already runs and cannot easily leave. There is rarely a separate negotiation, often little documentation, and no realistic option to walk away. Most community-bank AI arrives this way, and the honest fact is that the leverage to demand custom terms may not exist.

When it does not, governance shifts from negotiation to verification and documentation. Rely on the vendor's current SOC 2 Type II report—dated within the last twelve months—and any ISO/IEC 42001 certification, and contract for the right to receive the results even when direct audit rights are refused. Be clear about what those artifacts attest: a SOC 2 covers security, availability, and confidentiality controls—it says nothing about model performance, validation quality, or fair-lending outcomes—and ISO/IEC 42001 certifies a management system rather than tested results. They close the security question; the model-governance gap they leave is exactly what the documented risk acceptance has to own. For your largest core providers, the federal banking agencies examine the service provider directly, and those examination reports can supplement what you cannot extract by contract. Use user groups and shared assessments to apply collective pressure no single small bank can. Where a gap remains, document the compensating controls and the residual risk the bank has accepted—a deliberate, documented acceptance stands up to an examiner; silence reads as a gap nobody noticed.

Chapter Summary

This chapter has provided a comprehensive approach for governing vendor-supplied AI:

- Vendor AI governance is the bank's third-party risk-management life cycle—planning, due diligence, contract, monitoring, termination—with AI-specific content added at each stage.
- AI-specific due diligence evaluates vendor governance maturity, transparency, bias testing, data handling, and change management.
- Twenty-five critical questions provide a structured evaluation, with the full questionnaire in Appendix C.
- Eight Essential contractual provisions protect transparency, audit rights, change notification, and fair lending cooperation.
- Ongoing monitoring tracks vendor AI performance, outcomes, and governance throughout the relationship.
- Concentration risk assessment protects against excessive dependence on single AI vendors.
- Fourth-party and foundation-model concentration can hide a shared dependency behind apparent vendor diversity; require subprocessor visibility and a model-layer fallback.

Looking Ahead

Parts III and IV have assembled the governance program: structure, policies, risk assessment, and vendor governance. Two chapters remain in the risk picture—technical controls for AI systems (Chapter 10) and governance for AI that acts on its own (Chapter 11)—before Part V turns to oversight: the governance committee (Chapter 12), the board (Chapter 13), and internal audit (Chapter 14).

The Vendor Questionnaire

The 25-question Due Diligence Questionnaire in Appendix C covers the full scope of AI vendor evaluation. The AI Governance Toolkit provides it as an editable Excel workbook with a built-in SOC 2 review checklist, ready for your vendor management team to send as-is.

"The examiner never asked about our AI strategy. She asked who could turn the model off, and how fast."

— COMMUNITY BANK IT OFFICER —

✦ CHAPTER 10 ✦

Technical Controls for AI Systems

FROM THE STACK TO THE OFF-SWITCH

AI governance becomes real in the systems themselves. A policy the board approves and a risk the committee rates still have to be enforced somewhere concrete—configured in a platform, logged on a server, monitored in production, and ready to show an examiner. That enforcement is the technical layer: the controls a bank's technology function builds, runs, and defends. The model-risk discipline of SR 26-2, the FFIEC's IT examination booklets, and GLBA's information-security requirements—implemented for banks through the Interagency Guidelines Establishing Information Security Standards—already define what is expected of the systems a bank runs; this chapter is where those expectations become architecture, data controls, validation, monitoring, and security.

Two notes on scope. Community banks run this layer with far less than a large institution—tighter budgets, smaller teams, often one or two people carrying other duties as well—so every control here is calibrated to be implementable at that scale. And this is the technical layer only: building an AI-aware culture and training staff is Chapter 16, the governance committee is Chapter 12, and the fair-lending analysis behind a model's outcomes is Chapter 5. Where the technical layer hands off to one of them, the text says so.

The chapter follows the path an AI system takes through the bank: choosing the stack, readying the data, vetting the model, running it under watch, defending it, and managing the generative tools already in use. Set against NIST's AI Risk Management Framework, the rest of this book is Govern; this chapter is the controls of Map, Measure, and Manage—with ISO/IEC 42001 as the standard a bank can build the whole program against. Chapter 4 and Appendix L map the full landscape; this is where it meets the hardware. Each section ends in a concrete control, and the chapter summary gathers them into one place.

10.1 The AI Technology Stack

Most community banks will never build AI infrastructure from scratch; the stack is assembled from vendor selections and platform choices. The most consequential of those choices are made by default—a feature switched on in an existing system, a model reached through whatever cloud path a vendor happens to offer—and a choice made by default is a choice the bank let someone else make. Taking those decisions back is the IT leader's first job, because three of them shape the bank's exposure more than any policy will: where the AI runs, what data it can touch, and how fast it can be shut off.

In practice, a community bank chooses among four deployment models, and most end up running all four at once:

1. **Embedded AI:** software the bank already licenses—the core, the loan origination system, the fraud platform—turned on by a vendor release. The bank rarely chooses these deliberately; they arrive, which is why the inventory and vendor-notice discipline in Chapters 8 and 9 matters.

2. **Standalone SaaS AI:** a cloud product the bank buys and configures but does not host.

3. **Cloud AI in the bank's own tenant:** a foundation model reached through a private endpoint inside the bank's Azure, AWS, or Google environment, where the bank builds the application on top.

4. **Self-hosted models:** models running on the bank's own hardware, inside its perimeter.

Each model puts the bank's data and the point of control in a different place. For every AI system, that one fact—which model it runs on—dictates who can see the data, where inference happens, and how fast the bank can intervene.

DEPLOYMENT MODEL	WHERE IT RUNS	WHO HOLDS THE DATA	BANK'S CONTROL
Embedded AI	Inside software the bank already licenses	The vendor's environment	Limited—usually just on/off in the product
Standalone SaaS AI	The vendor's cloud	The vendor's cloud, under contract	Indirect—configuration and contract terms
Cloud AI in the bank's tenant	The bank's own cloud tenant	Stays in the bank's tenant	Direct—bank owns the endpoint, keys, and access
Self-hosted models	The bank's own hardware, on-premises	Entirely inside the bank's perimeter	Full—stop, roll back, or isolate in-house

CLOUD VS. ON-PREMISES

For most workloads, cloud delivery is appropriate and economical, and it meets OCC, FDIC, and FFIEC expectations when the vendor arrangement underneath is sound. The driver of the decision is data classification: the more sensitive the data a model touches, the more the bank has to know about where that data goes—which is a GLBA Safeguards question (where customer data is processed and stored, and who can reach it) before it is an AI one. Before a cloud AI arrangement carries customer data, five conditions must hold:

1. **Residency:** where the data is processed and stored.
2. **No training on the bank's data:** a contractual term, covered in Chapter 9, that the data will not be retained or used to train the vendor's models.
3. **Private connectivity:** encrypted connectivity that keeps the bank's traffic off the public internet.
4. **Identity and access:** controls that extend the bank's perimeter to the platform.
5. **A fallback:** a defined plan for when the service is unavailable.

One caution is worth acting on directly: most cloud contracts community banks signed were written for predictable transactional workloads and never anticipated AI's data movement or training-data questions, so an existing contract may not actually govern how a newly enabled AI feature uses the bank's data.

On-premises AI was until recently the rare exception, reserved for the most sensitive applications. The case for it is now stronger than most IT leaders assume. Capable open-weight models run on a single workstation-class GPU; once the hardware is in place, self-hosted inference costs a fraction of metered cloud APIs at steady volume; and the data never leaves the bank's walls, which removes an entire class of third-party risk. It carries real cost—hardware, model expertise, and the same monitoring, logging, and auditability the bank demands of any vendor—but for a bank whose most sensitive use case is the very thing blocking AI adoption, a small self-hosted deployment is now genuinely within reach.

INTEGRATION ARCHITECTURE

The most important architectural decision is what the AI is allowed to connect to. The durable pattern is to let AI systems read from a governed copy of the bank's data—the data warehouse or the core provider's data hub—instead of wiring them directly into the system of record. Reading is the lower-risk path; writing is where harm happens, so any route that lets an AI system post a decision or update a record belongs behind a human-approval gate with complete logging. The cores most community banks run—Fiserv, FIS, and Jack Henry serve roughly seventy percent of U.S. institutions—expose data and AI features through their own integration layers, so integration usually happens through those supported channels and the identities they authenticate; few community banks have cause to custom-wire a model into the core itself.

For each AI system, four things are worth documenting and keeping current: the data it takes in and how that data is classified; the connection and authentication method, including the service accounts or machine identities it uses, each held to least privilege; the outputs it produces and the systems they flow to; and the failure mode—what happens when the system is unavailable or returns a result the bank cannot rely on.

> **One Diagram, Four Jobs**
>
> Create a one-page architecture diagram for each High- and Critical-tier AI system showing data inputs and their classification, processing, outputs, integration points, and the off-switch. The FFIEC's Architecture, Infrastructure, and Operations booklet expects examiners to see exactly this kind of data-flow documentation, and it names AI explicitly. One diagram then does four jobs at once: it feeds the risk assessment, gives incident responders a map, demonstrates governance maturity to examiners, and lets people outside IT see what the system depends on.

10.2 Data Readiness and Data Quality

Data decides how reliable an AI system can ever be, and in a community bank the data is where AI governance is most often won or lost. Readiness—data that is accurate, complete, consistent, timely, accessible, and governed—is the largest barrier most banks face, and it sits more squarely with the IT leader than any other part of governance. No control downstream can rescue a model that is fed bad data. Most of this is GLBA Safeguards discipline the bank already owes—a data inventory and the controls around it—now pointed at the data that feeds a model. Data governance in full is its own discipline—it deserves its own book—and this section covers the part AI governance cannot do without.

ASSESSING DATA READINESS

Before any AI system goes live—or any vendor product is bought—its data has to clear a readiness assessment across the dimensions that decide whether a model can trust its inputs. For each one, the test is whether the data is good enough for the decision the AI will make:

- **Completeness:** Are the required fields populated, or are there gaps the model will silently fill with its own assumptions? Measure the missing-value rate for each field the system uses.
- **Accuracy:** Does the data match reality, verified against the source of record or a sample re-check, or is it carrying stale and mis-keyed values?
- **Consistency:** Does a customer, balance, or status read the same across the core, the loan origination system, and the CRM, or do the same fields disagree between systems?
- **Timeliness:** Is the data fresh enough for the decision? A fraud model on 24-hour-old data and a credit model on last year's income are deciding about a reality that has already moved.

- **Accessibility:** Can the data be reached programmatically through a supported interface, or does it require manual extraction that will not scale and cannot be audited?

- **Lineage:** Can the bank show where each field originates, how it moves, and what transforms it along the way? This is the question an examiner asks first.

- **Governance:** Does the data have an owner, an access policy, and change control, or is it managed informally?

For many community banks, an honest pass through these dimensions reveals real gaps. That is exactly why it is done before deployment: each gap is either remediated before launch or carried as a documented, compensating-control decision, so nothing the model depends on becomes a surprise in production.

DATA QUALITY MONITORING

Readiness is assessed once, at the gate. After go-live, feeds break and conditions move, so quality gets watched continuously. Automated checks at each AI system's input points track the same dimensions used in the readiness assessment—completeness, accuracy, timeliness, and consistency—against defined thresholds, and alert a named owner when a feed falls below them. A missing-value rate that jumps from two percent to fifteen, a nightly feed that quietly stops, or a field whose format changes after a vendor update are the everyday failures that degrade an AI system long before anyone notices a bad decision.

Watch as well for data drift: a gradual shift in the makeup of the incoming data away from what the model saw when it was built, driven by a new product, a changing customer mix, or a different rate environment. Drift on the input side is an early signal that the model's outputs are about to become less reliable, and it is the data companion to the model-performance monitoring in Section 10.5 and the drift detection covered in Chapter 8.

THE DATA READINESS SCORECARD

A model trained on incomplete data will produce incomplete results. A model fed inconsistent data will produce inconsistent decisions. A model operating on stale data will make decisions based on yesterday's reality. Score each system's data across the seven dimensions above before it goes live.

Score each dimension on a 1–4 scale. Any dimension scoring 1 must be remediated before the AI system goes live; a 2 requires the documented compensating controls described below. The overall readiness score is the lowest individual dimension score—the chain is only as strong as its weakest link. Appendix K provides a completed scorecard example. *(Editable version in the Toolkit)*

INTERPRETING THE SCORECARD

Overall score of 3–4: Data environment is ready for AI deployment. Proceed with normal governance processes. Document the assessment as part of the AI risk assessment.

Overall score of 2: Data gaps exist that could affect AI performance. Deploy AI only with documented compensating controls (such as enhanced output monitoring or human review) and a remediation plan with defined timelines.

Overall score of 1: Data environment is not ready for AI deployment. Remediate the identified dimension before proceeding. Deploying AI on unreliable data creates governance risk that downstream controls cannot adequately mitigate. A low score here is not a failed test. It is the cheapest warning the bank will ever get.

COMMON DATA READINESS FAILURES IN COMMUNITY BANKS

In practice, community banks most frequently struggle with three data readiness dimensions:

1. **Consistency across systems.** Customer data in the core banking system may not match customer data in the loan origination

system, the digital banking platform, or the CRM. When an AI system draws from multiple sources, these inconsistencies propagate into the model's inputs, producing unreliable outputs. The remediation is not necessarily a single master data management system—that may be a multi-year project. The immediate step is to identify the data sources each AI system uses, map the inconsistencies, and determine whether they are material to the AI's function.

2. **Timeliness of data feeds.** Many community bank data integrations run on batch schedules—nightly, weekly, or even monthly. Assess whether each AI system's data refresh frequency is appropriate for the decisions it makes.

3. **Data lineage documentation.** When an examiner asks, "Where does the data in this AI model come from?" the bank must be able to answer with specificity—not just the source system, but the extraction method, any transformations applied, and the path the data takes from source to model input. Community banks that have not documented data lineage should start with their High- and Critical-tier AI systems and work down.

> **The Data Readiness Investment**
>
> Data readiness work is not an AI-specific cost. Every dollar invested in data quality, consistency, and governance pays dividends across the entire organization—in more reliable reporting, faster regulatory response, better customer service, and more effective risk management. AI simply raises the stakes: it amplifies data quality problems into decision quality problems. Treat data readiness as an enterprise investment that happens to be essential for AI governance.

10.3 Model Development Life Cycle

Most community banks consume models built elsewhere; only some build or customize their own. For the IT leader, that makes the model development life cycle most useful as an oversight lens: it is the map of how a model comes to exist, and therefore the set of questions to put to a vendor and the stages to govern in any model the bank does build. A vendor who cannot speak clearly to each of these stages has told the bank something worth hearing. The sequence below is the model-risk discipline of SR 26-2 (the successor to SR 11-7) applied even to a model the bank did not build, and ISO/IEC 23894 frames the same arc as AI risk management. The life cycle has six stages.

1. **Problem formulation:** Defining the business problem, the success criteria, and the governance requirements before any data is touched. The test, in-house or with a vendor: can someone state in one sentence what the model decides, what good looks like, and what the model must never do? A model without that cannot be validated or monitored against anything.

2. **Data preparation:** Selecting, cleaning, transforming, and validating the training data, the stage where quality, bias, and representation problems are either caught or baked in. This is where the readiness work in Section 10.2 pays off, and where to ask a vendor what data the model was trained on and how closely it resembles the bank's own customers and markets.

3. **Model development:** Selecting the approach, training, and tuning. For community banks the right choice is usually a well-understood technique such as logistic regression or a decision tree that the bank can explain to an examiner and to a denied applicant; a more complex method needs lift the bank can actually justify. Simpler models are easier to validate, monitor, and defend.

4. **Testing and validation:** Independent evaluation of performance, including accuracy, stress, bias, and sensitivity testing, before the model is trusted. Section 10.4 treats this on its own; the principle

that matters here is that the people who validate a model are independent of the people who built it.

5. **Deployment:** Moving the validated model into production behind monitoring, logging, and a fallback. Treat it as a controlled change event, with a defined approval, a way to watch the first live decisions closely, and a tested path to roll back if it misbehaves.

6. **Monitoring and retirement:** Ongoing performance monitoring with defined triggers for revalidation, retraining, or retirement. Every model has a lifespan; it decays as data and conditions shift, and the bank needs both the criteria and the resolve to retire one that no longer earns its place. A model nobody is willing to switch off is a model nobody is truly governing.

10.4 Testing and Validation Techniques

A model the bank has never tested is a risk it cannot see. Validation is how that risk becomes visible—how the bank learns what a system actually does before it makes a decision the bank has to defend. For a community bank the real constraint is doing this meaningfully without a model-risk department, and most of it can be done from the outside, without ever seeing the vendor's code. The techniques below run from the most accessible to the most specialized. Independent validation, backtesting, and ongoing review are what SR 26-2 expects for a model in a regulated activity; the work here scales them to a community bank's resources.

OUTPUT ANALYSIS

The most accessible technique, and the one available even for a black-box vendor system, is systematic analysis of what the model produces. Track the distribution of outputs over time and watch for unexplained shifts: an approval rate that moves several points in a month, a fraud-alert volume that doubles, a chatbot whose escalation rate climbs. Compare the AI's output to human judgment on a regular sample—the rate at which a reviewer overturns the model is one of the most useful health signals a bank has. Break the outputs down by customer segment to surface disparities worth investigating, and pull the outliers where the result looks anomalous. None of this requires seeing inside the model, which is what makes it the backbone of ongoing validation for purchased AI.

BACKTESTING

Backtesting runs the model against historical data where the real outcome is already known, turning "we think it works" into a measured claim. For a credit model, compare its predictions to how the loans actually performed; for a fraud model, measure how many confirmed-fraud cases it would have caught and how many false alarms it would have raised along the way. The honest measures show the trade-off—catch rate against false-positive rate—because a single headline accuracy number can be flattered by a heavily imbalanced book. Regulators expect backtesting evidence for models used in regulated activities, so the results and their date belong in the model's file. For a newly signed vendor model, run it before go-live: score a sample of the bank's own past applications and compare the model's answers with the decisions the bank actually made—the cheapest preview of how the system will treat your market.

BIAS TESTING

Bias testing asks whether the system produces disparate outcomes for protected classes—comparing approval rates, pricing, and other results across groups, testing whether disparities are statistically significant or explained by legitimate factors, and checking for proxy features that stand in for protected characteristics. It matters most for lending models and should be considered for any system that touches customers. The IT leader's job here is to build the data and logging that make the test possible and to schedule it; the methodology itself, and the legal frame around it, belong to the fair-lending guidance in Section 5.2.

STRESS TESTING

Stress testing asks how the model behaves when conditions leave the range it was trained on. Run the credit model through a recession scenario and watch where its assumptions break; simulate a surge of novel fraud patterns and see how quickly the fraud model's performance falls off. The output that matters is the failure point—the conditions under which the bank should stop trusting the model and fall back to manual review—because that is what feeds the contingency plan and the human-fallback procedures noted in Section 10.1.

> **Resource-Constrained Validation**
>
> If a bank lacks in-house expertise for formal model validation, consider engaging a qualified third party. Several consulting firms specialize in model validation for community banks at price points appropriate for smaller institutions. As a rough benchmark, third-party model validation typically runs $15,000 to $50,000 per model depending on complexity, a dedicated bias testing engagement costs $10,000 to $30,000, and AI governance software platforms range from $30,000 to $100,000 per year. These numbers vary widely, but they give a starting point for budget conversations with the board. External validation costs a fraction of what an inadequate-model-risk-management finding costs—and unlike the finding, it can be budgeted. Alternatively, community bank networks and industry groups sometimes offer shared validation resources.

10.5 Monitoring, Logging, and Auditability

An AI system in production has to be watched and recorded. Every AI decision must be traceable—a practical requirement for regulatory compliance, incident investigation, customer dispute resolution, and internal audit—and the system's behavior has to be monitored closely enough to catch trouble before a customer or an examiner does. The IT leader owns both the watching and the record. Ongoing monitoring is the part of SR 26-2 banks most often let slip, and the audit trail is what the FFIEC's Audit booklet expects to find.

PERFORMANCE MONITORING

Validation proves a model works on the day it is tested; monitoring is how the bank knows it still works months later. For each High- and Critical-tier system, define the handful of metrics that signal health—for a credit model, approval rate and the override rate; for fraud, catch rate and false positives; for a chatbot, escalation and complaint rates—set thresholds on them, and route an alert to a named owner when a metric crosses the line. The triggers matter as much as the metrics: a threshold breach should start a review, and a sustained one should fire the revalidation, retraining, or retirement decision described in Section 10.3 and the drift detection in Chapter 8.

Monitoring also has to catch the AI incident that trips no traditional alarm. A firewall logs an intrusion, but nothing pages the IT leader when a vendor enables a new model, when an approval rate drifts five points over a quarter, or when an agent begins acting outside its intended scope. Those are visible only in the output and behavior metrics, which makes monitoring the bank's first and often only warning that something has changed.

LOGGING REQUIREMENTS

At a minimum, AI systems should log the following for each decision or output: the input data used, the output produced, a timestamp, the model version in use, and any confidence scores or uncertainty indicators. For High- and Critical-tier systems, logs should also capture the key factors driving each decision (reason codes), sufficient to support adverse action notices and customer explanations.

LOG RETENTION AND ACCESS

Log retention periods should align with the bank's record retention policy and applicable regulatory requirements. For lending decisions, this means retaining decision logs at least as long as the bank retains the underlying credit records—twenty-five months after the applicant is notified of the action taken, the Regulation B record-retention period for consumer credit—and longer where a fair-lending statute of limitations or a litigation hold extends that window. Logs must be stored securely with appropriate access controls and must remain accessible for audit, examination, and incident investigation.

AUDIT TRAIL DESIGN

Beyond individual decision logs, the broader governance life cycle has to be auditable too: model changes, validation results, monitoring reports, committee decisions, and incident responses, documented and retained so internal audit and examiners can reconstruct the governance narrative for any AI system at any point in time. These same artifacts—the architecture diagrams, the readiness scorecards, the monitoring dashboards—are what the IT leader carries into the governance committee and the board (Chapters 12 and 13). The technical record is also the governance record.

10.6 Security and AI-Specific Threats

AI systems face every threat a traditional system faces—malware, phishing, unauthorized access, denial of service—plus several threat vectors that exist only because the system learns and generates. The bank's security program has to reach to cover them. That reach is an extension of GLBA's information-security obligations—the Interagency Guidelines Establishing Information Security Standards—and the FFIEC's Information Security booklet to this new surface. The exposure is no hypothetical: in May 2026, a community bank holding company disclosed in an SEC filing a material incident involving nonpublic customer information shared with an AI tool.

KEY AI THREAT VECTORS

Data poisoning: An attacker manipulates the training data used by an AI model, causing the model to learn incorrect patterns. For community banks, the most likely data poisoning scenario involves a vendor's training data being compromised, potentially affecting all banks using that vendor's model.

Adversarial attacks: Deliberately crafted inputs designed to cause an AI model to produce incorrect outputs. For example, a fraudster might structure transactions in patterns designed to evade AI-based fraud detection.
Prompt injection: Specific to generative AI, prompt injection involves crafting inputs that cause the AI to bypass its safety controls or reveal confidential information. This threat is particularly relevant for customer-facing chatbots.
Model theft and extraction: Repeated queries to an AI system designed to reverse-engineer the model's decision logic, potentially revealing proprietary information or enabling adversaries to identify vulnerabilities.
Supply chain compromise: AI systems depend on complex software supply chains including libraries, frameworks, and pre-trained models. A compromise at any point in this chain can introduce vulnerabilities into the bank's AI systems.

GENERATIVE AI AND AGENT THREATS

The threats above come from classic machine learning. The assistants, chatbots, and agents a community bank actually deploys add several more, catalogued in the OWASP Top 10 for Large Language Model Applications and MITRE ATLAS—free, recognized references that work as the standing checklist for AI vendor diligence and the annual security assessment.

Insecure output handling: The AI's output is treated as trusted and passed straight into a web page, database query, or workflow, where it becomes an injection path into bank systems.

Sensitive-information disclosure: The model surfaces NPI or confidential data through its outputs, its logs, or the context it retrieves.

System-prompt leakage: An assistant's hidden instructions can be extracted, which makes the system prompt the wrong place for secrets or access rules; enforce those in the application and data layer.

Excessive agency: An agent is granted more tools, permissions, or autonomy than its task requires—the focus of the agentic controls later in this chapter and the Toolkit's Agentic-AI Guardrail Checklist.

Retrieval (RAG) data exposure: A retrieval-augmented chatbot—the 'ask the policy manual' or 'ask the loan file' assistant that is a common first build—answers by pulling documents into the model at query time. If retrieval does not enforce the same per-user permissions as the source system, the chatbot surfaces restricted or NPI documents to anyone allowed to use it. Test it directly: sign in as a low-privilege user and ask for something only a manager should see, and confirm with the vendor that retrieval is permission-aware.

Indirect prompt injection: Instructions hidden inside a retrieved document, email, or web page are obeyed by the model as though the user had typed them. Treat retrieved content as untrusted input.

SECURITY CONTROLS

Controls should map to the threats above and stay proportionate to the system's risk tier:

Untrusted inputs: Validate and sanitize what enters the model, and rate-limit and monitor query patterns to blunt adversarial probing and model-extraction attempts.

Untrusted outputs: Never pass model output straight into a database query, a web page, or a downstream action without the checks the bank applies to any external input. This closes the insecure-output-handling path.

Isolation and secrets: Put AI systems that handle sensitive data behind network segmentation, and hold secrets and access rules in the application and data layer, where they can be enforced.

Permission-aware retrieval: A retrieval-augmented assistant should surface only what the signed-in user is already entitled to see.

Supply-chain review: Extend vendor diligence to ask where models and their components come from and how the vendor secures its own training data and pipeline; a compromise upstream reaches every bank on that platform.

AI-specific testing: Add AI scenarios to the annual security assessment and penetration test, using the OWASP Top 10 for LLM Applications and MITRE ATLAS as the ready-made plan.

Agent controls: Scoped permissions, least privilege, and the autonomy caps and kill switch covered in Section 10.7.

None of this is exotic. It is the security discipline the bank already runs, told about a new kind of system.

CONTAINING AN AI INCIDENT

Decide how the bank will contain an AI incident before it has one. Because these failures often trip no traditional alarm, the response looks different from a classic breach: the IT leader should be able to take a model offline and fall back to manual processing, revoke an agent's credentials and access, roll back to a prior model version, and preserve the logs and prompts needed to reconstruct what happened. Who can pull that switch, and how fast, belongs in the incident response plan in Chapter 7, settled before the bank needs it.

10.7 Managing Generative AI Tools in the Enterprise

Generative AI is the most accessible AI in the bank and the least predictable, and that combination is what the IT leader has to manage. Any employee with a web browser can access a generative AI tool, and the outputs of these tools range from remarkably useful to dangerously wrong. The IT leader's role is to enable productive use while preventing harm. The exposure is a familiar one—customer NPI leaving the bank's control under GLBA—reaching new tools, most of them third parties the OCC's guidance already governs.

ENTERPRISE PLATFORM STRATEGY

The most effective approach is to provide employees with an approved, enterprise-grade generative AI platform that incorporates the bank's security and governance requirements. Enterprise versions of tools like ChatGPT, Microsoft Copilot, or similar platforms offer features that consumer versions do not: data handling agreements that prevent customer data from being used to train the vendor's models, administrative controls over who can access the platform and what they can do, audit logging of user interactions, content filtering and safety controls, and integration with the bank's identity and access management systems.

Give employees an approved platform and two problems shrink at once: their AI use moves into an environment the bank controls, and the consumer tools lose their appeal.

BEFORE ENABLING COPILOT: REMEDIATE OVERSHARING FIRST

An embedded assistant like Microsoft 365 Copilot or Google Workspace's Gemini inherits every permission that already exists in the environment. The moment it is enabled, it can surface any file a user can technically reach—including content exposed by broken permission inheritance, 'share with everyone' links, and over-permissioned sites. An employee can ask for the salary file or the board packet and receive it, with no attacker involved and no policy knowingly broken. Because this is a one-switch exposure of NPI, treat enabling an embedded assistant as a governed change with a short pre-enablement runbook: run the platform's permission and oversharing reports (Microsoft's SharePoint Advanced Management, included with the Copilot license, flags broken inheritance, broad-access sites, and anonymous links); apply a temporary containment such as Restricted SharePoint Search while the worst sites are remediated; apply sensitivity labels and data-loss-prevention rules so files marked confidential or containing NPI stay outside the assistant's reach; then enable the assistant for a pilot group, review the audit logs, and expand once the permission picture is clean. Every step uses tooling included in the Copilot license itself; the work is permission hygiene the bank should want regardless of AI.

TECHNICAL CONTROLS

Technology should reinforce policy. Four controls carry most of the weight: network-level blocks on known unapproved generative AI platforms; data-loss-prevention rules that detect and stop sensitive data from being pasted or uploaded into AI tools; endpoint or browser controls that show where generative AI is actually being used on bank devices; and disclosure scanning where policy requires AI-generated content to be flagged. Those same proxy, DLP, and SaaS-discovery logs are how the bank finds shadow AI technically—the unsanctioned tools employees adopt on their own—which then feeds the AI inventory and the discovery process in Chapter 8.

These controls should be implemented as guardrails. The goal is to prevent the most harmful behaviors—customer data leakage, unapproved tool use in regulated activities—while allowing employees to use generative AI productively within approved boundaries.

EMPLOYEE ENABLEMENT

Technical controls only hold if employees understand them. The IT leader's part is to put the guidance where the work happens—a prompt at the point of use about what data may and may not go into a tool, and an easy way to report a bad or surprising output. The training program itself, and the culture that makes people willing to disclose how they use AI, are the subject of Chapter 16. And the technical surface feeds the program: the shadow-AI discoveries, DLP hits, and override rates this chapter produces are the telemetry that tells Chapter 16 what to train on, and gives a blameless-reporting culture something concrete to act on.

AGENT IDENTITIES AND AUTONOMY CONTROLS

Agentic AI deserves the IT leader's specific attention. The first enterprise agents are arriving inside the productivity platforms banks already run, and they arrive with something traditional tools never had: their own identities, credentials, and standing access. Chapter 11 governs them in full—identity rules, autonomy caps, approval gates, and the kill switch—and the Toolkit's Agentic-AI Guardrail Checklist turns those requirements into a deployment gate.

> **Balancing Control and Enablement**
>
> The IT leader who bans all generative AI will not eliminate its use—employees will simply use it on personal devices, creating shadow AI that is completely outside the bank's governance program. The more effective approach is to provide approved tools, set clear boundaries, implement proportionate technical controls, and invest in training. Enable the behavior you want; make the behavior you don't want harder.

Chapter Summary

Choose the stack (10.1)

- ✦ Map each High- and Critical-tier system to one of the four deployment models, and record where it runs, who holds the data, and how it can be shut off.
- ✦ Before any cloud AI touches customer data, clear the five checks: residency, a no-training contract term, private connectivity, extended identity and access controls, and a defined fallback.
- ✦ Draw a one-page architecture and data-flow diagram for each High- and Critical-tier system.

Ready the data (10.2)

- ✦ Score each system's data on the seven readiness dimensions; remediate any dimension scoring 1 before go-live, and document compensating controls and a remediation timeline for any 2.
- ✦ Place automated data-quality checks at each system's input points, with thresholds and a named owner who is alerted when a feed slips.

Vet the model (10.3–10.4)

- ✦ For every model, obtain the six life-cycle artifacts—problem statement, training-data description, development choices, validation results, deployment plan, and retirement triggers—from the vendor or an in-house team.
- ✦ Run output analysis on every system; backtest and stress-test High- and Critical-tier models; schedule bias testing through Chapter 5 wherever customers are affected.

Run it under watch (10.5)

- Define health metrics, thresholds, and triggers for each system; alert a named owner, and tie a sustained breach to revalidation, retraining, or retirement.
- Log the inputs, output, model version, timestamp, and reason codes for each decision; retain decision logs for at least the Regulation B period (25 months for consumer credit) and longer under a litigation hold.

Defend it (10.6)

- Apply the seven security controls, from treating inputs and outputs as untrusted to permission-aware retrieval and tight agent controls.
- Write the AI incident-containment plan before it is needed: who can take a model offline, revoke an agent's access, roll back a version, and preserve the logs.

Manage the tools everyone is using (10.7)

- Stand up an approved enterprise generative-AI platform, and remediate oversharing before enabling an embedded assistant like Copilot or Gemini.
- Discover shadow AI through proxy, DLP, and SaaS-discovery logs, and feed what it surfaces into the AI inventory (Chapter 8).
- Govern agent identities with least privilege, logging, autonomy caps, and a tested kill switch.

WHERE TO START

The book already maps the first quarter. The Quick Start—Your First 90 Days of AI Governance—sets six program actions: discover, classify, govern, adopt the policies, assess, and report. Within the technical layer, three controls are the foundation the rest depend on, and they come first when resources are tight:

1. **Build the inventory:** the Quick Start's own first move, and for good reason—a system cannot be governed, secured, or explained until it is written down, and every later step depends on it (Section 10.1; the discovery method is in Chapter 8).

2. **Draw the data boundary:** one enforceable rule—what customer data may never enter an unapproved tool—backed by DLP and shadow-AI discovery, which stops the most common and most damaging failure while the rest of the program is still being built (Sections 10.2 and 10.7).

3. **Gate the decisions that can hurt a customer:** credit denials, fraud holds, anything that moves money—a human in front, logged, until the bank has earned confidence in the system (Sections 10.1 and 10.5).

Do these three and the program the Quick Start lays out is anchored in the systems themselves. The rest of this chapter's checklist is how it matures from there.

Looking Ahead

Chapter 11 takes the agent controls introduced in Section 10.7 and builds the full governance frame around them—the decisions that always require a person, agent identity and access, and the records that let you reconstruct what an agent did and why. Chapter 12 then shifts audience to the governance committee, covering how to make it a working body—meeting structures, dashboards, challenge frameworks, and decision criteria.

Scoring Data Readiness

The Data Readiness Scorecard in Appendix K gives you the methodology. The AI Governance Toolkit provides it as an editable Excel workbook your IT team can complete in one session—with built-in scoring and a ready / conditional / not-ready readout.

"The chatbot says something wrong and a person catches it before it matters. The agent does something wrong and you find out when the customer calls."

— COMMUNITY BANK INFORMATION SECURITY OFFICER —

✦ CHAPTER 11 ✦

Governing Agentic AI

WHEN AI STOPS ANSWERING AND STARTS ACTING

The governance most community banks are building was designed for AI that answers a question. The AI arriving next takes a sequence of actions. An agentic system can pull a customer's history, draft a response, send it through the CRM, schedule a follow-up, and log the interaction—five steps a person used to own, now run in sequence by software. These systems are already in production at institutions not much larger than yours, and they change the control problem. This chapter sets out how to govern AI that acts, before it acts at scale.

Consider how the failure mode differs. A deposit-operations agent reads a returned-mail flag, extracts a new address from an uploaded document, updates the customer record, reissues two statements to the new address, emails a confirmation, and logs the interaction—six steps, no waiting. The extraction was wrong: the document belonged to a different customer. Nobody looks, because nothing failed. The error surfaces when a customer calls about a move he never made—after statements carrying another customer's account history entered the mail stream under his name, and after a confirmation email announced the change. No single step was reckless. The chain was the failure—and the disclosure obligations are already in motion. Case Study 1 at the end of the book tells a fuller version of this story; this chapter builds the controls that keep it fiction.

11.1 What Changes When AI Takes Action

A conversational AI produces an output for a person to use. An agentic AI takes the next step itself, and then the step after that. The risk lives in the chain: each action becomes the input to the next, so a small early error compounds, and the point at which a human could have caught it has already passed. Three questions decide whether an agentic system is governable: what it is allowed to do without a person, what it can reach, and how fast it can be stopped. The rest of this chapter answers them.

The risk assessment you already built knows what to do with agents. In Chapter 8's scoring language, agentic capability drives the Decision Autonomy dimension to its top scores—a system that executes on its own judgment scores a 4 nearly by definition—and under the highest-single-dimension rule, that places agentic deployments in Tier 3 or Tier 4 governance before any other dimension is considered. Pre-deployment assessment, enhanced monitoring, and committee approval are the default for any agentic deployment.

11.2 Human-in-the-Loop: The Decisions That Always Need a Person

Some workflows can run end to end without a person; requiring sign-off at every step would erase the benefit. But certain decisions should require human authorization regardless of how confident the system appears. At a community bank that list almost always includes credit decisioning, suspicious activity reporting, wire transfer authorization, customer data disclosure, account restrictions for suspected fraud, and any regulatory filing. Decide that list now, write it into your AI governance policy, and build it into the system as a hard gate. Defining the bright lines before deployment is far cheaper than retrofitting them after an agent has done something no one authorized.

DECISION	WHY IT STAYS HUMAN	REGULATORY ANCHOR
Credit decisions —approve, deny, or price	Adverse action duties and fair lending exposure attach to the decision	ECOA / Regulation B; Fair Housing Act
SAR filing decisions	Confidentiality rules and investigative judgment; the filing decision is the bank's legal act	BSA; FinCEN guidance
Wire transfer release	Irrevocable movement of funds; fraud exposure concentrates here	UCC Article 4A loss-allocation regime; bank wire policy
Customer data disclosure to third parties	Privacy obligations and reputational stakes	GLBA / Regulation P
Regulatory filings and certifications	Accuracy attestations carry institutional and personal accountability	Agency filing requirements
Account restriction or closure for suspected fraud	Customer harm and complaint exposure require judgment	UDAAP considerations

A bright line is real only if the system cannot route around it. In practice, that means the gate lives in the platform's permission model rather than in the prompt: the agent's credentials lack the entitlement to execute the gated action, and the workflow routes to a named human queue. A gate implemented as an instruction—"always ask before sending a wire"—is a request, and an agent under load treats requests as suggestions. Test the gate the way you test the kill switch: attempt the gated action in a sandbox and confirm the system refuses—and logs the refusal.

11.3 Agent Identity, Access, and the Kill Switch

An agent that acts needs an identity, credentials, and access. Treat an agent identity the way you treat a service account: least privilege, scoped permissions, no standing access it does not need, and activity logging on everything it touches. Review its entitlements on the same calendar as your human access reviews—an agent's permissions drift the same way people's do. Give every agentic deployment an autonomy cap (the value, volume, or risk level above which it must hand off to a person), an approval gate on the decisions in Section 11.2, and a tested kill switch that an operator can use to stop it immediately. A kill switch that has never been tested is not a control.

11.4 Contracting for Agents

Most community banks will meet their first agent inside a vendor platform, which makes Chapter 9's contract toolkit the natural home for agent-specific provisions. Five belong in any agreement that includes agentic capability:

1. Autonomy disclosure. The vendor specifies exactly what the agent may do without human action—and any expansion of that scope requires advance written notice and the bank's opt-in. Autonomy never widens through a routine release.

2. Action logs as a deliverable. Complete, exportable logs of every agent action, retained on terms the bank controls rather than merely viewable in the vendor's console.

3. Kill-switch service levels. A documented mechanism to suspend the agent immediately, with a tested procedure, a named support path, and a committed response time.

4. Liability allocation. Explicit treatment of responsibility for autonomous actions, including errors that propagate before detection.

5. Sandbox and test rights. The bank may attempt gated actions in a non-production environment to validate the controls in Section 11.2.

Appendix C's questionnaire covers vendor AI generally; these five provisions extend it for the agentic case, and the questions in Section 9.2 about model change notification apply with double force when the model can act.

11.5 Monitoring AI That Acts

The monitoring disciplines of Section 10.5 carry over to agents, with five agent-specific metrics added to the dashboard:

1. Escalation rate—the share of tasks the agent hands to a human. A falling rate can mean improving quality; it can also mean drift in what the agent treats as routine. Investigate the trend in either direction.

2. Autonomy-cap hits—how often the agent reaches its value, volume, or risk ceiling. Frequent hits mean the cap is working, and they predict pressure to raise it. Raising a cap is a governance decision for the committee, documented like any new use case.

3. Override rate at approval gates—how often humans reject the agent's recommendation. A near-zero rate sustained for months means the gate has stopped gating; sample those approvals in the next audit.

4. Kill-switch test cadence—time since the stop mechanism was last exercised. Quarterly is a reasonable default for Tier 3 deployments; monthly for Tier 4.

5. Actions per completed task—the average number of steps the agent takes to finish its work. A sudden rise flags an agent looping or improvising outside its lane.

These five lines belong on the Chapter 12 risk dashboard for every deployed agent, alongside the standard performance and drift metrics.

11.6 Records, Audit, and Accountability

When an agent acts, it creates evidence and obligations. The prompts that drove it, the steps it took, and the outputs it produced are business records, with retention, litigation-hold, and audit-trail implications—and most community-bank records programs were written before any of this existed. Log each agent action with enough detail to reconstruct the sequence and answer one question for any decision: why did the system do that? Accountability does not transfer to the software. A named human owner remains responsible for what the agent does, the same as for any other system the bank runs.

Agentic AI is early, but it is arriving on a vendor's release schedule, faster than most strategic plans anticipate. The institutions that decide their bright lines, identity rules, and kill-switch discipline now will adopt it on their own terms. Chapter 17 returns to where this is heading; the task in this chapter is to be ready before the first agent is switched on.

Chapter Summary

This chapter has built the governance frame for AI that acts:

- Agentic AI changes the control problem: risk lives in the chain of actions, where an early error compounds past the point where a person could have caught it.
- Three questions decide governability—what the agent may do without a person, what it can reach, and how fast it can be stopped.
- Bright-line decisions—credit, SAR filings, wires, data disclosure, account restrictions, regulatory filings—require hard gates built into the platform's permissions, with the gate itself tested in a sandbox.
- Agent identities are service accounts: least privilege, scoped access, full activity logging, and an autonomy cap set before deployment.
- Vendor contracts need agent-specific provisions: autonomy disclosure, exportable action logs, kill-switch service levels, liability allocation, and sandbox rights.
- Five dashboard metrics keep an agent honest—escalation rate, autonomy-cap hits, override rate at gates, actions per task, and kill-switch test cadence.
- Under Chapter 8's methodology, agentic capability scores Tier 3 or Tier 4 by default; committee approval and enhanced monitoring are the baseline.

Looking Ahead

Part V turns from systems to oversight. Chapter 12 makes the governance committee a working body—meeting rhythms, dashboards, effective challenge, and decision criteria—and the agent metrics above give that committee some of its most consequential new lines to watch.

Guardrails Before Go-Live

The Agentic-AI Guardrail Checklist in the AI Governance Toolkit turns this chapter's controls—bright lines, identity rules, autonomy caps, approval gates, kill-switch tests, vendor contract provisions, and the monitoring metrics that follow go-live—into a pre-deployment sign-off your governance committee can require for any agent.

PART V

Oversight and Assurance

"The difference between a governance committee that works and one that doesn't comes down to whether anyone in the room is willing to ask the uncomfortable question."

— COMMUNITY BANK BOARD CHAIR —

✦ CHAPTER 12 ✦

The AI Governance Committee

MAKING THE COMMITTEE A WORKING BODY

The risk facing any governance committee is ceremonialism: meetings that follow a rote agenda, reports that are received but not questioned, approvals that are granted without meaningful challenge. A committee can hold every meeting on its charter's calendar and still provide no oversight at all.

Chapter 6 established the committee's charter, composition, and place in the organizational structure. This chapter makes the committee operational, providing the practical tools—meeting structures, dashboards, challenge frameworks, decision criteria, and review processes—that prevent that outcome and turn a committee on paper into one that provides genuine oversight.

12.1 Committee Meeting Structure and Cadence

Community bank leaders are busy, and meetings that consume entire afternoons will quickly lose attendance and engagement. The following structure is designed for a quarterly meeting of approximately two hours, with room to stretch when an agenda item needs it.

MODEL QUARTERLY MEETING AGENDA

AGENDA ITEM	TIME	PRESENTER	ACTION
Opening & Prior Meeting Action Items	10 min	Committee Chair	Review
AI Inventory Update – New/ Changed Systems	15 min	IT Leader	Awareness
Risk Dashboard Review	15 min	Risk/Compliance	Discussion
High/Critical Tier Monitoring Reports	20 min	System Owners	Challenge
New AI Use Case Proposals	20 min	Business Line Sponsor	Approve / Reject
Vendor AI Updates & Due Diligence Findings	10 min	IT Leader/Vendor Mgr	Awareness
Incident Review & Lessons Learned	10 min	Incident Owner	Discussion
Regulatory & Policy Updates	10 min	Compliance	Discussion
Items for Board Reporting/ Escalation	10 min	Committee Chair	Decision

ANNUAL PLANNING CALENDAR

Beyond the standard quarterly agenda, certain governance activities run on an annual cycle—four items, one per quarter, an hour of preparation each, and the annual review writes itself from them:

Q1: Annual AI governance policy review. Assess whether policies remain current and effective. Approve updates.

Q2: Annual AI risk appetite review. Confirm or revise the bank's risk appetite for AI, ensuring alignment with the strategic plan.

Q3: Annual AI inventory completeness review. Commission a fresh discovery exercise to identify any AI systems that may have entered the environment without governance.

Q4: Annual governance effectiveness self-assessment (Section 12.6). Evaluate whether the committee is fulfilling its charter obligations and identify what to improve. Prepare the annual governance report for the board.

MEETING MATERIALS AND PRE-READING

Meeting materials—the risk dashboard, monitoring reports, use case proposals, and any other agenda items—should be distributed at least five business days before the meeting. Materials should include a brief executive summary highlighting items requiring decision or escalation, so members can focus their preparation on the most consequential issues.

12.2 The AI Risk Dashboard

The AI risk dashboard is the committee's primary tool for understanding the bank's AI risk posture at a glance.

DASHBOARD COMPONENTS

An effective AI risk dashboard for a community bank should include the following elements:

- **AI inventory summary:** Total number of AI systems by risk tier, with counts of new systems added and systems retired since the last report.

- **Risk tier distribution:** A visual representation showing the proportion of systems at each risk tier, enabling the committee to assess the overall risk profile.

- **Monitoring status:** For each High- and Critical-tier system, a status indicator showing whether the system is performing within acceptable parameters (green), showing signs of concern warranting investigation (yellow), or exhibiting problems requiring immediate attention (red).

- **Incident summary:** Count and severity of AI-related incidents since the last report, with status of any open incidents.

- **Vendor AI summary:** Status of vendor AI relationships, including any material changes, due diligence findings, or contractual concerns.

- **Agent metrics:** For any deployed agentic system, the five lines from Section 11.5—escalation rate, autonomy-cap hits, gate override rate, actions per task, and kill-switch test currency.

- **Overdue actions:** Any governance actions—risk assessments, validations, monitoring reviews, policy updates—that are past their scheduled completion date.

- **Regulatory developments:** Brief summary of regulatory changes or examination findings relevant to AI governance.

The dashboard should fit on one to two pages. If it is longer, it contains too much operational detail for committee-level consumption. Detailed supporting data should be available on request but should not be included in the default presentation.

> **Design Principle**
>
> The dashboard should answer four questions in under five minutes: What AI systems do we have and how risky are they? Are they performing as expected? Has anything gone wrong since the last meeting? Are there any governance gaps or overdue actions? If the dashboard answers these four questions clearly, it is doing its job.

SAMPLE DASHBOARD: WHAT IT LOOKS LIKE

Below is a simplified example of the High- and Critical-tier status section. Your dashboard should carry at least this level of detail for every system that warrants committee attention. Appendix D shows a completed quarterly board report illustrating how this information integrates into the full reporting structure.

SYSTEM	RISK TIER	STATUS	LAST REVIEW	KEY NOTES / NEXT ACTION
Credit Scoring Model	Critical	Yellow	Mar 2026	Fair lending analysis in progress; target June 2026
Fraud Detection	Critical	Green	Feb 2026	False positive rate 12.3% (target <15%)
BSA/AML Alerts	High	Green	Feb 2026	Annual validation complete; no findings
Digital Chatbot	High	Red	Jan 2026	Scope creep detected; guardrails added; re-review due Apr

Status: Green = performing within parameters. Yellow = concern under investigation. Red = requires immediate attention.

12.3 Effective Challenge: Asking the Right Questions

The governance committee's most important function is effective challenge—the practice of constructively questioning assumptions, probing risk assessments, and holding management accountable for governance commitments. Effective challenge is what distinguishes a governance committee from a rubber stamp.

Effective challenge requires both institutional support and individual skill. The committee's charter must authorize challenge, the meeting culture must welcome it, and individual members must be equipped with the right questions. Here are frameworks for challenging the most common committee agenda items.

CHALLENGING RISK ASSESSMENTS

- What assumptions underlie this risk classification? What would change if those assumptions were wrong?
- How was bias testing conducted, and what were the specific results?
- What is the worst-case scenario if this AI system fails, and what is our fallback plan?
- Has this system been tested with data from our specific customer base, or only with the vendor's general test data?
- Who validated this assessment, and are they independent from the team that developed or selected the system?

CHALLENGING MONITORING REPORTS

- Are the KPIs we are monitoring actually the right ones? Would a different metric tell us something important that we are currently missing?
- This system has been green for six quarters—are we sure our monitoring is sensitive enough to detect emerging problems?

- What would have to change in the business environment for this system's risk tier to increase?
- Are we monitoring outcomes for all customer segments, or only in aggregate?

CHALLENGING NEW USE CASE PROPOSALS

- What problem does this AI system solve that cannot be solved by existing non-AI methods?
- What customer data will this system process, and have customers consented to this use?
- How does this proposal align with the bank's stated AI risk appetite?
- What is the exit strategy if this system does not perform as expected or if the vendor relationship deteriorates?
- Has the compliance team confirmed that this use case does not trigger additional regulatory requirements?

12.4 Approving AI Use Cases: Decision Criteria and Documentation

The committee's approval decisions should be based on standardized criteria. Standardization ensures consistency, reduces the influence of personality and politics, and creates a defensible decision record for examiners. These criteria are the committee's operational test for an individual use case; the board-level Investment Decision Matrix in Section 13.3 is the strategic screen for major AI commitments. The two work in sequence—the board's screen comes first for significant investments, the committee's test before anything goes live.

DECISION CRITERIA

For each AI use case presented for approval, the committee should evaluate five criteria:

1. **Business justification:** Is the proposed AI use case aligned with the bank's strategic objectives? Does it solve a real business problem? Is the expected benefit proportionate to the cost and complexity?

2. **Risk assessment adequacy:** Has a thorough risk assessment been completed, proportionate to the proposed system's risk tier? Are all five risk dimensions evaluated with supporting evidence?

3. **Control adequacy:** Are the proposed controls—monitoring, validation, human oversight, incident response—sufficient to manage the identified risks?

4. **Regulatory alignment:** Does the proposed use case comply with all applicable regulatory requirements? Has the compliance function confirmed alignment?

5. **Resource readiness:** Does the bank have the technical infrastructure, data readiness, staffing, and budget to support the proposed AI system through its full life cycle, including ongoing monitoring and governance?

The committee's decision—approve, approve with conditions, defer pending additional information, or reject—should be documented in the meeting minutes along with the rationale. Conditional approvals should specify the conditions that must be met and the timeline for meeting them. A conditional approval is a deadline, and examiners treat an expired one as a finding.

12.5 Managing AI Incidents and Escalations

When an AI incident occurs, the committee's role shifts from oversight to active governance. The committee does not manage the incident operationally—that is the responsibility of the IT leader and the incident response team—but it ensures that the response is adequate, that root causes are identified, and that lessons are incorporated into the governance program.

THE COMMITTEE'S INCIDENT RESPONSIBILITIES

When an escalation reaches the committee, it does four things:

1. confirms the severity call and that the right response level is running;
2. asks whether the incident exposed a gap in the program itself;
3. decides whether the board needs to hear about it—and whether regulators do;
4. commissions a post-incident review to find the root cause and what to fix.

Post-incident reviews are among the most valuable governance activities. They should examine what went wrong technically and whether the governance program—risk assessment, monitoring, controls, escalation procedures—functioned as intended. Findings should be translated into specific governance improvements with assigned owners and deadlines.

Culture of Learning

The committee sets the tone for how the organization responds to AI failures. If failures are met with blame and punishment, employees will hide problems and avoid reporting. If failures are treated as learning opportunities—while still maintaining accountability—employees will report early, escalate honestly, and contribute to continuous improvement. The committee sets this tone in its first incident review—everyone watches what happens to the first person who brings bad news.

12.6 Annual AI Governance Review

Once each year, the committee should step back from operational oversight and assess the effectiveness of the governance program itself. This annual review serves multiple purposes: it identifies governance gaps before examiners do, it ensures the program remains aligned with the bank's evolving AI use, and it demonstrates governance maturity to the board and regulators.

REVIEW SCOPE

The annual governance review should evaluate:

- Whether AI governance policies remain current and effective.
- Whether the risk assessment methodology is capturing the right risks at the right intensity.
- Whether monitoring and validation activities are being performed as scheduled and producing meaningful results.
- Whether the committee is fulfilling its charter obligations with appropriate rigor.

- Whether reporting to the board provides adequate information for fiduciary oversight.

- Whether the bank's AI risk profile has changed materially and whether governance has adapted accordingly.

- Whether training and awareness programs are reaching all relevant employees.

REVIEW OUTPUT

The annual review should produce a written governance effectiveness report that is presented to the board. The report should include an overall assessment of governance maturity—using the maturity roadmap from Chapter 17 (Foundational, Evolving, Established, and Advanced)—along with specific findings, improvement recommendations, and an action plan for the coming year. When an examiner asks how governance has improved year over year, this report is the answer the bank hands across the table.

Chapter Summary

This chapter has provided the practical tools for making the governance committee operational:

- A structured quarterly meeting agenda with pre-distributed materials ensures committee meetings are productive and focused.
- The AI risk dashboard provides a concise, actionable view of the bank's AI risk posture on one to two pages.
- Effective challenge frameworks equip committee members with specific questions for risk assessments, monitoring reports, and use case proposals.
- Standardized decision criteria and documentation create consistent, defensible approval decisions.
- Incident management and post-incident review close the loop between operational failures and governance improvement.
- The annual governance review ensures the program itself remains effective and demonstrates maturity to the board and examiners.

Looking Ahead

Chapter 13 turns to the board of directors—the third audience this book serves directly, after the IT leaders of Chapters 10 and 11 and the governance committee of this chapter. Its focus: fiduciary oversight, reporting, strategic direction, and the questions every board member should ask.

The First Board Report

The AI Risk Dashboard and Board Reporting in Appendix D give you the reporting structure. The AI Governance Toolkit provides them in the Board AI Reporting Pack—an editable PowerPoint deck with Word and Excel companions—populate it with your data and the first committee presentation is drafted.

"My job as a director is to understand the risk. The algorithm has its own experts."

— COMMUNITY BANK BOARD DIRECTOR —

✦ CHAPTER 13 ✦

The Board's Role in AI Governance

FIDUCIARY OVERSIGHT WITHOUT MICROMANAGEMENT

The board of directors bears ultimate fiduciary responsibility for the bank's risk posture, and that responsibility now extends unambiguously to AI risk. Yet most community bank board members did not enter their roles expecting to oversee artificial intelligence, and many feel uncertain about what effective AI oversight looks like.

This chapter is written for board members. It lays out what the board should do about AI governance and what it should leave to management. It names the information the board needs, gives a working approach to strategic oversight, addresses the fiduciary and liability stakes, and ends with the questions every board member should be ready to ask.

13.1 What Boards Need to Know (and What They Don't)

The most common concern board members express about AI governance is some version of: *I don't understand this technology well enough to oversee it.* This concern is understandable but misplaced. Board members oversee many complex domains—interest rate risk, cybersecurity, regulatory compliance—without being technical experts. The same model applies to AI.

WHAT BOARDS NEED TO KNOW

- ✔ **What AI systems the bank uses:** The board should understand the inventory of AI systems, their risk classifications, and which business functions they support. This does not require understanding the algorithms; it requires understanding the landscape.

- ✔ **How AI risk is governed:** The board should understand the governance structure—committee composition, policies in place, risk assessment methodology, monitoring cadence—and be confident that it is functioning effectively.

- ✔ **What the significant risks are:** The board should understand the bank's most material AI risks and what management is doing to mitigate them.

- ✔ **How AI aligns with strategy:** The board should understand how AI use supports the bank's strategic objectives and whether the bank's AI risk appetite is appropriate.

- ✔ **What could go wrong:** The board should understand the plausible worst-case scenarios and the bank's preparedness for them.

WHAT BOARDS DO NOT NEED TO KNOW

Boards do not need to understand:

- ✕ How machine learning algorithms work
- ✕ What specific parameters or features a model uses
- ✕ The technical details of model validation
- ✕ How data pipelines are designed or monitored
- ✕ The vendor's proprietary methodology

These are management responsibilities. The board's role is to ensure that management has established appropriate processes to address these areas.

> **The Governance vs. Management Boundary**
> The clearest test for whether a question belongs to the board or to management: If the question can only be answered by someone with technical expertise, it is a management question. If the question can be answered in business risk language, it is a board question. "What algorithm does our fraud model use?" is a management question. "What happens if our fraud model fails?" is a board question.

13.2 Board-Level AI Risk Reporting

The quality of AI governance at the board level depends entirely on the quality of information the board receives. Too little information, and the board cannot fulfill its oversight role. Too much information—or information presented in technical language—and the board is overwhelmed rather than informed.

RECOMMENDED REPORTING FORMAT

The board should receive a quarterly AI governance report from the governance committee. The report should be concise—no more than three to five pages—and structured to answer the board's core questions. A recommended format includes:

1. An executive summary of the bank's AI risk posture in two to three sentences
2. The AI inventory summary showing system counts by risk tier with changes since last report
3. A status overview of High- and Critical-tier systems including performance and any concerns
4. An incident summary covering any AI incidents with status and lessons learned
5. Governance activity highlights such as policies reviewed, assessments completed, and validations performed
6. Regulatory developments relevant to the bank
7. Items requiring board action or awareness

REPORTING FREQUENCY

Quarterly reporting is the recommended baseline for most community banks. Banks with extensive AI use or rapidly evolving AI portfolios may benefit from more frequent reporting. Between quarterly reports, the board should receive immediate notification of material AI incidents or regulatory developments, as defined in the escalation protocols established in Chapter 6.

The board should resist the temptation to request more frequent routine reporting unless the bank's AI risk profile genuinely warrants it.

Over-reporting creates administrative burden without improving governance quality and can dilute the significance of the reports the board receives.

13.3 Strategic AI Oversight

Beyond risk oversight, the board plays a strategic role in AI governance. This strategic role involves three activities: setting the bank's AI risk appetite, keeping AI use aligned with the bank's mission, and drawing the ethical boundaries that strategy must respect.

AI RISK APPETITE

The board should articulate the bank's appetite for AI risk—the level of AI-related risk the bank is willing to accept in pursuit of its strategic objectives. The risk appetite statement need not be technical or quantitative. It should express the bank's philosophy in practical terms: which types of AI use the bank is comfortable pursuing, which types require exceptional justification, and which types are outside the bank's tolerance. For example, a board might approve moderate-risk AI applications that enhance operational efficiency while placing AI that makes autonomous lending decisions without human review outside its tolerance altogether—the bright line Chapter 11 already draws.

The AI Governance Toolkit provides an AI Risk Appetite Statement template the board can adopt and review annually.

MISSION ALIGNMENT

Community banks exist to serve their communities, and the board should ensure that AI adoption reinforces rather than undermines this mission. Questions of mission alignment include whether AI is being used to enhance customer relationships or to replace them, whether AI-driven efficiency gains are being reinvested in the community or merely extracted as cost savings, and whether AI systems treat all community members equitably, particularly historically underserved populations. These are values questions, and the board is uniquely positioned to address them.

ETHICAL BOUNDARIES

The board should establish clear ethical boundaries for AI use that reflect the bank's values. These boundaries go beyond regulatory compliance to address questions that the law does not yet resolve: Is it appropriate to use AI to analyze customer social media for creditworthiness signals? Should the bank use AI that makes recommendations its own staff cannot explain? At what point does AI-driven personalization cross into surveillance? The board need not answer every ethical question in advance; its job is to set the tone and authorize management to bring the hard ones forward.

Prioritizing AI Investments

Beyond setting risk appetite and ethical boundaries, the board should ensure that the bank has a coherent strategy for AI adoption—one that distinguishes between AI investments that are urgent, strategic, and exploratory, and aligns AI spending with the bank's broader strategic plan. The following approach helps boards evaluate AI investment decisions without requiring technical expertise.

THE THREE HORIZONS OF AI ADOPTION

Horizon 1—Defend: AI that the bank must adopt to maintain regulatory compliance, competitive parity, or operational viability. Examples include AI-powered fraud detection (fast becoming table stakes across the industry), BSA/AML transaction monitoring (moving from rules-based to AI-driven across the industry), and basic cybersecurity tools that incorporate AI threat detection. Horizon 1 investments are mandatory in practice—the open question is how to govern them. The board's role is to ensure these systems are deployed with appropriate governance and adequate resources.
Horizon 2—Differentiate: AI that could provide meaningful competitive advantage or operational improvement, but is not yet essential. Examples

include AI-assisted lending that improves approval speed and credit quality, digital banking chatbots that extend service hours without extending staffing, generative AI productivity tools that improve employee efficiency, and customer analytics that enable more relevant product recommendations. Horizon 2 investments require a business case: What is the expected benefit, what is the cost, what are the risks, and does the benefit justify the governance investment? The board should evaluate these on a case-by-case basis.

Horizon 3—Explore: AI capabilities that are emerging but not yet mature enough for community bank deployment. Examples include agentic AI that could autonomously execute multi-step banking processes, AI-driven real-time loan pricing optimization, and voice-based AI banking assistants. Horizon 3 requires awareness today; investment decisions can wait. The board should ask management to provide a brief annual update on emerging AI trends (Chapter 17 provides the structure) so the bank is not caught off guard when Horizon 3 technologies move to Horizon 2. One caveat keeps Horizon 3 honest: agentic capabilities are already arriving embedded in the platforms banks run today, so while the deployment decision can wait, the governance posture—block, monitor, or pilot—cannot.

THE AI INVESTMENT DECISION MATRIX

When a specific AI investment is proposed—whether by management, a vendor, or a board member who read about it at a conference—the board should evaluate it against five criteria:

CRITERION	KEY QUESTION	GREEN (FAVORS PROCEEDING)	RED (FAVORS WAITING)
Strategic Fit	Does this AI use case support our strategic plan?	Directly supports a stated strategic priority	No clear connection to strategy; technology for its own sake
Customer Impact	Will this improve the customer experience?	Clear customer benefit (speed, access, accuracy)	Benefit is purely internal; no customer value or potential customer harm

CRITERION	KEY QUESTION	GREEN (FAVORS PROCEEDING)	RED (FAVORS WAITING)
Proportionate Risk	Is the governance cost proportionate to the benefit?	Governance requirements are manageable with existing resources	Would require significant new staff, systems, or budget to govern properly
Competitive Position	What happens if we don't do this?	Competitors already offering this; we risk losing customers	No competitive urgency; this can wait without consequence
Reversibility	Can we undo this if it doesn't work?	Clear exit strategy; can revert to manual processes	Deep integration; high switching costs; vendor lock-in

What this matrix produces is a structured conversation. The board's value lies in asking the questions the matrix contains and insisting on clear answers from management before approving significant AI investments.

WHEN TO SAY "NOT YET"

One of the board's most important strategic contributions is the willingness to slow down. The AI industry generates enormous pressure to adopt quickly—vendors pitch urgency, conferences create fear of missing out, and media coverage implies that every bank must be an AI leader. The board should be the counterweight to this pressure, asking whether the bank has the governance maturity to manage the proposed AI system effectively. If the honest answer is no, the responsible decision is to invest in governance first and adopt the AI system when the foundation is ready. A bank that deploys AI it cannot govern is simply accumulating risk.

> **Strategic Patience**
> Community banks win by being smart movers rather than first movers. Speed without governance is just velocity toward risk.

13.4 Fiduciary Duties and Personal Liability

Board members are often concerned about personal liability related to AI governance. While the legal landscape is still developing, several principles are clear.

THE BUSINESS JUDGMENT RULE

Under the business judgment rule, board members are generally protected from personal liability for business decisions made in good faith, on an informed basis, and in the honest belief that the decision is in the bank's best interest. For AI governance, this means that a board that establishes a reasonable governance program, receives and reviews regular reports, asks appropriate questions, and acts on material concerns is generally well protected, even if an AI system subsequently produces harmful outcomes. The complement to that protection is the duty of oversight: under the Caremark line of cases, an utter failure to implement a reporting system for a known, mission-critical risk—or a conscious failure to monitor one—is the narrow path around the business judgment rule. It is one more reason the documented AI reporting of Section 13.2 matters.

WHAT CREATES LIABILITY EXPOSURE

Conversely, liability exposure increases when the board is unaware of the bank's AI use because it never asked, when the board receives warning signs—audit findings, incident reports, examiner concerns—and fails to act, when the board fails to establish any governance program for AI risk, or when the board approves AI use in regulated activities without understanding the associated risks. The governance program in this book is designed to provide the informed, engaged, documented oversight that the business judgment rule requires.

Liability Protection in Practice

The single most important thing a board can do to manage personal liability related to AI is to ensure that its governance activities are documented. Minutes that reflect substantive discussion, questions asked and answered, decisions made with rationale, and follow-up actions tracked are the evidence of informed engagement the business judgment rule protects. When someone hostile reads your minutes years later, engaged oversight should be the only available conclusion.

13.5 Board Education and Development

AI literacy at the board level is built over years, in increments, through a sustained and practical education program—and no board member needs to become a technologist along the way.

EDUCATION APPROACHES

- **Dedicated AI briefings:** Schedule one or two board education sessions per year focused on AI topics. These sessions might feature the bank's IT leader explaining current AI use, an external speaker providing industry context, or a tabletop exercise simulating an AI governance scenario.

- **Conference and industry resources:** Encourage board members to attend community banking conferences that address AI governance, and circulate relevant articles and industry reports between meetings.

- **External advisors:** Consider engaging an external AI governance advisor who can provide the board with independent perspective and expertise. This is particularly valuable during the initial years of AI governance development.

- **Board composition:** Over time, consider whether the board's composition includes members with technology or risk management backgrounds who can provide deeper expertise. That rarely means adding a "technology director" tomorrow; it means weighing technology and risk backgrounds in long-term succession planning.

13.6 Questions Every Board Member Should Ask

This section provides a concise reference of high-impact questions that board members can use during any meeting where AI is discussed. Each one does real work: it shows management the board is paying attention, it pulls a material risk into the open, or it pins an answer to a name and a date. Appendix H provides these in a one-page reference card format, along with a bonus set of vendor and resilience questions.

ABOUT THE AI LANDSCAPE

- How many AI systems are we using today, and how does that compare to a year ago?
- Which AI systems have the highest risk classification, and why?
- Are there any AI systems operating outside our governance program?

ABOUT RISK AND PERFORMANCE

- Have any AI systems experienced performance problems or produced unexpected results since our last meeting?
- What are our most significant AI risks today, and what is management doing to address them?
- Has any AI system been tested for bias in the last twelve months? What were the results?

ABOUT GOVERNANCE EFFECTIVENESS

- Is our governance program keeping pace with our AI adoption? Are there gaps?
- Are governance activities—assessments, validations, monitoring—being completed on schedule?
- What did our last examination reveal about AI governance, and have we addressed all findings?

ABOUT STRATEGY AND ETHICS

- How is AI supporting our strategic plan and our mission to serve this community?
- Are there AI applications we should be pursuing that we are not? Are there applications we should reconsider?
- Are we confident that our AI systems are treating all customers fairly?

Chapter Summary

This chapter has established how boards can provide effective AI oversight without technical expertise:

- Boards need to understand the AI landscape, governance program, significant risks, strategic alignment, and worst-case scenarios.
- Quarterly reporting of three to five pages, supplemented by immediate escalation of material events, provides appropriate board-level information.
- Strategic oversight includes setting AI risk appetite, ensuring mission alignment, and establishing ethical boundaries.
- Documented governance engagement protects board members under the business judgment rule.
- Sustained education—briefings, conferences, advisors, and board composition—builds AI literacy incrementally.
- A focused set of high-impact questions enables every board member to provide effective oversight immediately.

Looking Ahead

Chapter 14 completes the oversight picture with the third line of defense—internal audit's evolving role in AI governance, including the AI audit universe, audit methodology, AI-assisted auditing, and communicating findings so they get acted on.

Hand This to Your Board Members

The Board Member AI Question Reference Card in Appendix H fits on a single page. The AI Governance Toolkit includes it in the Board AI Reporting Pack—print copies for every board member before your next AI governance discussion.

"Audit's real question in AI governance is whether the governance system itself works as designed."

— COMMUNITY BANK CHIEF AUDIT EXECUTIVE —

✦ CHAPTER 14 ✦

Internal Audit and AI

THE THIRD LINE'S EVOLVING ROLE

Internal audit occupies a unique position in AI governance. As the third line of defense, audit provides the independent assurance that the governance program—the structure, policies, processes, and controls described in this book—is functioning as intended. Audit does not own AI systems, does not develop policies, and does not manage risk. Audit evaluates whether the people and processes responsible for these functions are doing their jobs effectively.

For community banks, where internal audit is often outsourced or performed by a small internal team with broad responsibilities, AI audit must be practical, focused, and proportionate. This chapter provides the methodology to make it so.

THE PROFESSIONAL STANDARDS BEHIND AI AUDIT

Internal audit does not approach AI freehand. The work is governed by the IIA's Global Internal Audit Standards, effective January 9, 2025, which organize the profession around five domains and fifteen guiding principles—among them risk-based planning, demonstrated competency, and a functional reporting line from the chief audit executive to the board. Those principles apply to AI exactly as they apply to any other auditable area.

For AI specifically, the IIA's Artificial Intelligence Auditing Framework (2024) provides the supplemental map. It is built around three components—AI strategy, governance, and the human factor—and includes a practitioner's checklist a small audit function can use as a starting point. It is guidance rather than a mandate: as of this writing, there is no AI-specific Topical Requirement. The mandatory Topical Requirements that do touch AI are the Cybersecurity requirement and the Third-Party requirement (effective September 2026), which apply wherever an AI system is also a cybersecurity or third-party matter—which is to say, almost always.

14.1 Developing an AI Audit Universe

The audit universe for AI encompasses all auditable activities related to AI governance. It is broader than the AI inventory itself, taking in the governance processes that surround the systems along with the systems themselves.

EIGHT COMPONENTS OF THE AI AUDIT UNIVERSE

1. **AI governance program:** The adequacy and effectiveness of the governance structure, committee operations, policies, and reporting processes.

2. **AI inventory completeness:** Whether the AI inventory accurately captures all AI systems in the organization, including vendor-embedded AI.

3. **Risk assessment process:** Whether risk assessments are being conducted as required, at the appropriate depth, with adequate documentation.

4. **Individual AI systems:** For High- and Critical-tier systems, whether specific governance requirements—validation, monitoring, bias testing, documentation—are being met.

5. **Vendor AI governance:** Whether vendor due diligence, contractual provisions, and ongoing monitoring are adequate.

6. **Policy compliance:** Whether employees are adhering to AI governance policies, particularly the acceptable use and generative AI policies.

7. **Training and awareness:** Whether training programs are being delivered, attended, and effective.

8. **Incident response:** Whether AI incidents have been identified, escalated, investigated, and remediated in accordance with policy.

RISK-BASED AUDIT PLANNING

Not every element of the audit universe needs to be audited every year. Audit planning should be risk based, with frequency calibrated to the risk and maturity of each area. Audit High- and Critical-tier AI systems every year. Assess the governance program itself annually while the program matures. Put lower-risk areas—minimal-tier systems, policy compliance for well-established processes—on a longer cycle. And get to newly deployed AI systems and recently changed governance processes in their first year, while the habits are still forming.

A three-year cycle keeps the universe covered without overwhelming a small audit function:

AUDIT AREA	FREQUENCY	NOTES
Governance program and committee effectiveness	Annual	Charter, minutes, challenge quality, decision documentation
AI inventory completeness	Annual	Independent discovery reconciled to the official inventory
High- and Critical-tier systems	Annual, rotating sample	Full governance testing for each sampled system
Vendor AI governance	Year one, then every two years	Due diligence, contract provisions, ongoing monitoring
Policy compliance, including acceptable use	Every two years	Documentation review plus behavioral testing
Training, awareness, and incident response	Every two to three years	Delivery, attendance, effectiveness; incident handling

14.2 AI Audit Methodology

Auditing AI governance requires a methodology that evaluates both the governance process and the governed system. The audit should assess design effectiveness (are the controls designed appropriately?) and operating effectiveness (are the controls actually working as designed?).

WHAT AUDIT DOES NOT DO

The most common mistake in AI audit is asking internal audit to do the second line's job. Audit does not re-validate models, re-run bias tests, or review model code. Its role is to provide independent assurance that validation happened, that it was performed by someone independent and competent, that it exercised genuine challenge, and that its findings were acted upon. The shorthand is useful: audit validates the validator.

This boundary matters more under SR 26-2, the model risk management guidance that replaced SR 11-7. SR 26-2 judges validation by the quality of the review itself, giving banks more flexibility in who performs it, and it leaves generative and agentic AI outside its scope. For those systems, independent audit assurance is often the most concrete oversight a bank has—which raises, rather than lowers, the bar for getting the audit right.

GOVERNANCE PROCESS AUDIT

The governance process audit evaluates the infrastructure of governance. Key testing procedures include reviewing committee minutes for evidence of substantive discussion, effective challenge, and timely decision-making. The auditor should examine risk assessments for completeness, consistency, and appropriate classification. Policy documents should be reviewed for currency, board approval, and alignment with actual practices. Reporting should be assessed for accuracy, completeness, and timeliness at each level—from operational monitoring to board reporting. The auditor should verify that escalation protocols have been followed when triggering events occurred.

AI SYSTEM AUDIT

For individual AI systems, particularly High- and Critical-tier systems, the audit should evaluate documentation adequacy—whether the system's purpose, inputs, logic, outputs, limitations, and risk assessment are documented sufficiently to support governance and examination. The test is whether a competent stranger could pick up the file and understand what the system does and why the bank trusts it. Assess validation currency: has independent validation been performed as required, and have findings been addressed? Test monitoring effectiveness: are KPIs being tracked at the required cadence, and are deterioration trends being identified and investigated? For vendor-supplied systems, the auditor should assess due diligence completeness, contractual provision adequacy, and ongoing monitoring practices.

COMPLIANCE TESTING

The audit should test compliance with AI governance policies through a combination of document review, interviews, and substantive testing. For the acceptable use policy, this might involve reviewing network logs for unauthorized AI tool access, interviewing employees about their AI usage, or testing whether data loss prevention controls are functioning. For the generative AI policy, testing might involve reviewing a sample of AI-assisted outputs to assess whether human review occurred as required.

AUDITING THE GOVERNANCE COMMITTEE ITSELF

One audit area deserves explicit mention because it is politically delicate: the governance committee. The committee's minutes, the quality of its challenge, and the documentation behind its approvals are auditable artifacts—and at a community bank, the committee usually includes the auditor's own senior colleagues, sometimes the CEO. The independence framing matters here. The chief audit executive's functional reporting line to the board's audit committee exists precisely for this situation, and the test is straightforward: do the minutes show real questions asked and answered, do approvals reference the documented criteria of Section 12.4, and were any decisions made outside the charter's authority? An audit function that cannot ask those questions about its own executives is signaling the answer.

14.3 Your First AI Audit

A bank that has never audited AI should resist the urge to audit everything. The first engagement has two jobs: establish whether the governance foundation exists, and create the baseline that future audits will measure against. Scope it accordingly—governance program, inventory completeness, and one high-tier system.

Begin with an evidence request to management. Seven items reveal most of what year one needs to know:

1. **The AI inventory, with risk tiers and the date each entry was last reviewed.**
2. **The governance committee charter and the last four sets of minutes.**
3. **The adopted AI policies, with board or committee approval dates.**
4. **The most recent risk assessment for one High- or Critical-tier system, with its approval.**
5. **The vendor due diligence file for that same system.**
6. **The most recent board AI report and any management responses to it.**
7. **The AI incident log—or management's written confirmation that no AI incidents occurred in the period.**

What the evidence shows determines the posture. Missing items are themselves the finding—an absent inventory or an unapproved policy is a clearer message than any nuanced testing result. Where the foundation exists, apply the illustrative program below to the selected system and write the baseline.

Illustrative Audit Program:
Governance of a High- or Critical-Tier AI System

A methodology is easier to trust when you can see it applied. The following is a short, illustrative program for auditing the governance of one high- or critical-tier AI system. The full work program is more extensive; this shows the shape of it.

1. **Inventory and discovery.** Independently look for AI in use—through network and SaaS-proxy logs, expense records, and a brief business-unit survey—and reconcile what you find to the official AI inventory. Pass: every system you discover is inventoried and tiered. Fail: any "shadow" AI, or a critical system mis-tiered as lower risk.
2. **Pre-deployment risk assessment.** For a sample of systems put into use during the period, request the risk assessment and its approval. Pass: the assessment is dated before go-live and approved by the authorized body. Fail: missing, back-dated, or unapproved.
3. **Validation independence and currency.** Request the validation report and the validator's reporting line. Pass: validation was performed independently of development and use, within the required cycle, with findings tracked to closure. Fail: the developer validated their own work, the validation is stale, or findings remain open.
4. **Monitoring against thresholds.** Pull the monitoring record for the period. Pass: performance metrics were tracked at the required cadence and threshold breaches were escalated. Fail: gaps in monitoring, or breaches with no escalation.
5. **Acceptable-use and generative-AI compliance.** Test that controls block unauthorized tools, and sample generative-AI-assisted work for evidence of the required human review. Pass: data-loss controls function and human review is documented. Fail: customer or confidential data can leave, or outputs were used unreviewed.
6. **Agentic guardrails, where applicable.** Inspect the logs for human-in-the-loop approval gates above the defined risk threshold, and confirm a kill switch exists and has been tested. Pass: gates fire and the kill switch works. Fail: no approval record, or an untested kill switch.

7. **Reporting integrity.** Trace one quarter's AI metrics from their source to the board package. Pass: accurate and complete. Fail: material omissions.

Sampling: for critical-tier systems, test them all; for populations of routine items, attribute sampling is sufficient.

When Audit Is Outsourced

If your bank outsources internal audit, ensure that the audit firm has AI governance competence. Many traditional audit firms are still developing this expertise. Ask the firm to describe their AI audit methodology, identify the team members who will perform the work and their qualifications, and provide references from other community bank clients where they have performed AI governance audits. If the firm cannot show that expertise, supplement the engagement with a specialist.

Year two expands coverage using the cycle table in Section 14.1; year three completes the universe. By the third year, the work has folded into the ordinary audit plan.

14.4 Using AI to Audit AI

Audit can put AI to work on its own files—and the moment it does, it takes on every obligation it tests the rest of the bank against, whether the function is in-house or outsourced.

POTENTIAL AI APPLICATIONS IN AUDIT

Four uses earn their keep in a small audit shop. Automated analysis runs through large data sets and flags the anomalies and patterns worth investigating. Natural language processing reads policy documents, contracts, and meeting minutes for completeness and consistency. Continuous monitoring tools give the auditor real-time assurance on AI system performance. And generative AI helps draft workpapers, findings documentation, and reports—drafts the auditor then verifies line by line.

GOVERNANCE OF AI AUDIT TOOLS

If audit uses AI tools, those tools should be subject to the same governance program as any other AI system in the bank. This includes inventory listing, risk assessment, data handling controls, and output verification. The credibility of audit findings depends on the reliability of audit tools, and an audit tool that produces inaccurate results undermines the entire assurance function.

A practical consideration: if audit uses generative AI to draft findings or recommendations, every output must be independently verified for accuracy before inclusion in audit reports. Audit reports carry significant weight with management, the board, and regulators; AI-generated errors in audit reports would be particularly damaging.

14.5 Communicating Audit Findings

AI audit findings are valuable only if they drive action. The auditor's challenge is translating technical observations into findings that management, the governance committee, and the board can understand and act upon.

WRITING EFFECTIVE AI AUDIT FINDINGS

Each finding should follow the standard audit finding structure—condition, criteria, cause, effect, and recommendation—but with attention to the audience. The condition should describe what the auditor observed in factual, non-technical language. The criteria should reference the specific policy, procedure, or regulatory expectation that was not met. The cause should explain why the deficiency exists, distinguishing between design failures and operating failures. The effect should articulate the risk to the bank in business terms: "model AUC declined" becomes "the fraud detection system may be missing more fraudulent transactions, increasing the bank's financial loss exposure and regulatory risk." The recommendation should be specific, actionable, and proportionate.

RATING AND PRIORITIZATION

Findings should be rated by severity to help management and the committee prioritize remediation. A simple three-level scale—high, moderate, low—is sufficient for most community banks. High findings indicate material governance deficiencies that require immediate management attention and board awareness. Moderate findings indicate meaningful gaps that should be addressed within a defined time frame. Low findings identify improvement opportunities that enhance governance maturity.

FOLLOW-UP AND TRACKING

The audit function should track remediation of all findings and report the status to the audit committee and board. Management should provide remediation plans with specific actions, responsible parties, and target completion dates. The governance committee should receive notification of all high and moderate AI governance audit findings and monitor remediation progress.

Continuous Auditing, Independence, and the Examination Payoff

Two terms are easy to confuse. Continuous monitoring is management's job—the first and second lines watching their own controls in near-real time. Continuous auditing is audit's own independent testing, on its own evidence. Audit can draw on management's monitoring, but it cannot let that monitoring stand in for its own work.

Independence is what makes the work credible. Under the Global Internal Audit Standards, the chief audit executive reports functionally to the audit committee of the board and confirms the function's independence in writing each year. There is now a direct payoff for getting this right: under the Federal Reserve's revised supervisory operating principles, examiners will rely on a satisfactorily-rated internal audit function to validate the remediation of supervisory findings rather than re-performing that work themselves. A credible AI audit function satisfies the standards and, increasingly, lightens the bank's examination burden.

Chapter Summary

This chapter has outlined internal audit's evolving role in AI governance:

- The AI audit universe spans governance processes, individual AI systems, vendor oversight, policy compliance, training, and incident response.
- Risk-based audit planning ensures annual coverage of high-risk areas while cycling through lower-risk areas on longer schedules.
- Audit methodology combines governance process evaluation, individual system assessment, and compliance testing.
- AI-assisted auditing can enhance efficiency but must be governed with the same discipline applied to other AI systems.
- Findings should be communicated in business language with clear severity ratings, actionable recommendations, and tracked remediation.
- Professional standards anchor the work—the IIA Global Internal Audit Standards and AI Auditing Framework—with audit assuring the validation function rather than re-validating models.
- A worked, step-by-step audit program makes the methodology concrete, and a credible audit function earns examiner reliance instead of re-examination.

Looking Ahead

Part VI turns from oversight to application. Chapter 15 brings the governance program to life across the AI use cases most relevant to community banking—from lending and fraud detection to generative AI productivity tools, and the threat side of the ledger: deepfakes, voice cloning, and synthetic identity fraud.

Give Your Auditors a Head Start

The AI Governance Toolkit includes the AI Internal Audit Work Program—aligned with the methodology in this chapter—plus an Excel companion with the regulatory mapping, evidence binder, and testing workpapers, ready for your next audit cycle.

PART VI

Putting Governance to Work

"Every governance program looks good on paper until your vendor updates a lending model on a Friday afternoon and your loan officers are using it Monday morning."

— COMMUNITY BANK IT OFFICER —

✦ CHAPTER 15 ✦

AI Use Cases Across the Bank

GOVERNANCE APPLIED TO REAL SCENARIOS

The preceding chapters built the governance program: principles, regulations, structure, policies, risk assessment, vendor management, audience-specific playbooks, and ethical practices. This chapter brings the program to life by applying it to the eight AI use cases most common in community banking—and to the AI-enabled fraud now aimed at banks and their customers.

Each section follows a consistent structure: a description of the use case, the typical risk classification, the primary governance concerns, the specific governance actions required, and practical recommendations. The goal is for the program to stop feeling like doctrine and start feeling like a regular Tuesday morning.

USE CASE	TYPICAL TIER	PRIMARY RISK	KEY GOVERNANCE	SECTION
AI Credit Scoring	Critical	Fair lending	Bias testing, adverse action	15.1
Fraud / BSA Monitoring	High	Effectiveness, false positives	Tuning, human oversight	15.2
Customer Chatbot	Moderate–High	Inaccuracy, UDAAP	Output monitoring, disclosure	15.3
Back-Office RPA	Minimal–Moderate	Operational	Inventory, periodic review	15.4
Employee generative AI	Moderate	Data leakage, accuracy	Acceptable use, DLP	15.5
Marketing Analytics	Moderate–High	Fair lending, privacy	Segmentation review, opt-out	15.6
AI in Compliance & Regulatory Monitoring	Moderate	Over-reliance; false confidence in outputs	Control testing, human review, documented limitations	15.7
AI-Enhanced Cybersecurity	Moderate–High	Deepfakes, adversarial inputs, alert fatigue	Out-of-band verification, tuning, human oversight	15.8, 15.9
AI-Enabled Fraud (threat)	n/a—threat	Synthetic identity, deepfakes, APP fraud	Synthetic-resistant verification, training	15.9

15.1 AI in Lending and Credit Decisions

AI-driven lending is the highest-stakes AI use case in community banking. A credit decision lands on someone's kitchen table—and it lands inside the most heavily regulated, most litigated corner of banking, where a model's mistake becomes an enforcement action with the bank's name on it. Most community banks encounter AI in lending through vendor-supplied credit scoring models, loan origination platforms with AI-powered decisioning, and pricing engines that use algorithmic methods.

GOVERNANCE REQUIREMENTS

Given the critical risk classification, lending AI gets the full treatment, and the work divides into five disciplines the bank already knows:

1. Document the model—inputs, logic, performance—and make the vendor supply what the bank cannot see.

2. Validate independently before deployment and keep the validation current: revalidation when the model materially changes, currency reviewed at least annually.

3. Test for fair lending on the Chapter 8 cadence—quarterly outcome monitoring for Critical-tier models, statistical disparate-impact testing on a window sized to the bank's application volume—with methodology and results written down.

4. Explain every denial with specific, meaningful reasons drawn from the model's actual decision factors.

5. Watch the model in production—accuracy, stability, outcome distributions across customer segments—with escalation triggers defined in advance.

Banks using automated valuation models in covered mortgage decisions have one more rule on point—the interagency AVM quality-control rule finalized in

2024 under Dodd-Frank Section 1125, which requires documented controls for confidence in estimates, protection against data manipulation, conflict-of-interest avoidance, random sample testing, and nondiscrimination compliance. For HMDA reporters, one more mirror: the bank's reported mortgage data is public, and examiners will compare what the model produces with what the bank reports—so any AI feature that influences a HMDA-reportable data point belongs in the model's documentation.

THE HUMAN REVIEW IMPERATIVE

Community banks should maintain meaningful human review in the lending process, particularly for credit denials and borderline decisions. "Meaningful" means that the human reviewer has the authority, training, and information to override the model's recommendation when circumstances warrant. The loan officer who reviews an AI credit recommendation should understand the key factors driving the recommendation and have the institutional support to exercise independent judgment.

Adverse Action Notices for AI-Driven Lending

When an AI model drives a credit denial, the bank must provide the applicant with an adverse action notice that identifies the principal reasons for the denial. Regulation B requires that these reasons be specific and indicate the principal factors that contributed to the adverse decision. ECOA is only half the notice obligation. When the denial rests in whole or in part on information from a consumer report, the Fair Credit Reporting Act adds its own duties—identifying the bureau that supplied the report—and when a credit score is used in the decision, the bank must disclose the score, its range, and the key factors that lowered it. An AI model that consumes bureau data triggers both statutes, and the notice has to satisfy each. Two more FCRA corners deserve a line in the bank's procedures: adverse action based on third-party information that is not a consumer report carries its own disclosure duty on the applicant's request, and risk-based pricing notices apply when bureau data leads to materially less favorable terms.

The fundamental problem is translation. The model's internal logic is arithmetic—it weights variables, calculates probabilities, produces a score. The notice has to turn that arithmetic into reasons a person can read, understand, and act on.

What Compliance Requires

Specificity: "Your credit application did not meet our criteria" is not acceptable. "Your debt-to-income ratio of 48% exceeded our maximum threshold of 43%" is specific and actionable.

Accuracy: The reasons listed must actually reflect the factors that drove the denial. If the model denied the application primarily because of insufficient credit history, the notice must say so.

Actionability: Ideally, the applicant should be able to understand from the notice what they could change to improve the outcome in a future application.

Governance Requirements for Adverse Action

The governance requirements for AI-driven adverse action notices should include three specific controls.

1. Reason code mapping: the bank must maintain a documented mapping between the AI model's internal decision factors and the consumer-facing reason codes used in adverse action notices. This mapping must be reviewed whenever the model is updated.

2. Reason code clarity review: have a non-technical reviewer—ideally someone from customer service or the branch network—read the reason codes and confirm they are understandable to a typical applicant.

3. Reason code coverage: ensure that the AI model's reason code library covers the full range of factors the model actually uses in denial decisions. A model that considers fifty variables but maps to only four reason codes is not meeting the specificity standard.

> **The Compliance Officer's Test**
> Pull ten recent AI-driven credit denials at random. Read each adverse action notice. For each one, ask: Could I explain this reason to the applicant in a face-to-face conversation? Does the reason accurately reflect why the model recommended denial? Could the applicant reasonably understand what to change to improve their outcome? If the answer to any question is no, the adverse action process needs remediation.

> **Vendor Negotiation Point**
> When evaluating a vendor's lending AI, the single most important capability to negotiate is adverse action reason code generation. If the vendor's model cannot produce specific, accurate reason codes that comply with Regulation B, the bank faces an immediate compliance risk. This capability should be a non-negotiable requirement in vendor selection.

15.2 AI in Fraud Detection and BSA/AML

Fraud detection and BSA/AML transaction monitoring are among the most mature AI applications in banking. The fit is natural: a monitoring model reads the entire transaction stream and surfaces patterns no human reviewer would see. The governance stakes are just as concrete. Tuning decisions decide what the model catches and what it misses, and the consequences of getting that balance wrong run from regulatory criticism to criminal liability.

GOVERNANCE REQUIREMENTS

Five disciplines keep a monitoring model honest. The governance program should address model tuning and threshold management with documented rationale for all tuning decisions, regular effectiveness reviews comparing the model's alerts to confirmed fraud or suspicious activity, false positive analysis to ensure the alert volume is manageable and that compliance staff can provide meaningful review, human decision authority for SAR filing ensuring that AI informs but does not make the SAR decision, and scenario testing to evaluate the model's performance against emerging fraud typologies and money laundering methods.

THE FALSE POSITIVE CHALLENGE

AI-powered transaction monitoring systems can generate enormous alert volumes. If the false positive rate is too high, compliance staff become desensitized, genuine suspicious activity is missed in the noise, and the bank incurs unnecessary investigation costs. The governance committee should monitor false positive rates as a key performance indicator and ensure that tuning decisions are documented and defensible. Examiners frequently ask about false positive management as an indicator of governance maturity.

15.3 AI-Powered Customer Service

Customer-facing AI—chatbots, virtual assistants, intelligent routing—creates a governance challenge none of the other use cases share: it talks to your customers, unscripted, in real time. Four things go wrong. The chatbot gets a fact wrong—a rate, a fee, a policy. It says something that crosses the UDAAP line. It discloses customer information it should have protected, or collects information nobody consented to. Or it fails the customers who need the most patience—people with disabilities or limited English.

GOVERNANCE REQUIREMENTS

Four controls keep a customer-facing AI inside its lane. Governance should include clear disclosure that the customer is interacting with an AI system rather than a human, with an easy path to reach a human agent. Output monitoring should involve regular review of chatbot conversations—both automated and manual sampling—to identify inaccurate, inappropriate, or potentially harmful responses. Escalation rules should define topics and customer situations that the chatbot should automatically hand off to a human, including complaints, complex product questions, and any indication of customer distress. Regular content review ensures that the chatbot's knowledge base remains accurate as products, rates, and policies change.

> **Start with a Sample**
> Start by reviewing a random sample of chatbot conversations monthly. Look for factual errors, inappropriate responses, customer frustration indicators, and privacy concerns. This simple monitoring practice catches problems early and provides evidence of governance for examiners. As your confidence grows, you can implement more sophisticated automated monitoring.

15.4 AI in Back-Office Operations

Back-office AI—document processing, data extraction, reconciliation, report generation, and workflow automation—typically carries lower risk than customer-facing or lending applications. Deposit operations are a natural starting point: automated exception processing, intelligent document handling for account opening, and AI-assisted daily reconciliation are among the lowest-risk AI applications a community bank can deploy, and they build institutional comfort with AI governance before you move to higher-stakes applications like lending or compliance.

These systems usually process internal data, assist rather than replace human workers, and do not make decisions that directly affect customers. Governance for back-office AI should be proportionate to this lower risk profile.

GOVERNANCE REQUIREMENTS

For minimal-risk back-office AI, governance requirements are streamlined: include the system in the AI inventory, assign an owner, document a brief description of what the system does, and conduct an annual review to confirm continued appropriate performance and risk classification.

For moderate-risk back-office AI—systems that process financial data, produce regulatory reports, or make decisions that indirectly affect customers—add a formal risk assessment, designated monitoring with periodic accuracy checks, and committee awareness.

The key governance question for back-office AI is whether the system's outputs are verified before they are used in downstream processes. An RPA system that generates a reconciliation report should have its output checked against source data on a sample basis. An AI system that extracts data from loan documents should have its extraction accuracy validated before that data is used for underwriting. The verification cadence should match the risk of downstream consequences if the AI produces an error.

WHEN BACK-OFFICE AI FAILS SILENTLY

The risk unique to back-office AI is silent failure. A customer-facing chatbot that gives bad advice generates an immediate complaint. A back-office AI system that misclassifies documents, miscalculates reconciliation totals, or extracts incorrect data from loan files can operate undetected for weeks or months—because no customer sees the output directly. By the time the error surfaces, it has already contaminated downstream processes.

Consider a common scenario: a vendor updates the layout of a standard loan document, and the bank's AI-powered data extraction tool begins pulling income figures from the wrong field. The system does not flag an error—it confidently extracts the wrong number. For three weeks, loan files move through underwriting with incorrect income data before a loan officer catches the discrepancy during a manual review. The bank must now re-underwrite every affected file, notify borrowers whose decisions may have been impacted, and explain the situation to examiners who discover the gap during the next examination.

This scenario illustrates why back-office AI governance is not optional simply because the risk tier is lower. The governance controls are simpler, but they must exist—and the most critical control is output verification.

MONITORING KPIS FOR BACK-OFFICE AI

Even minimal- and moderate-risk back-office AI systems should be tracked with a small set of key performance indicators. You do not need a complex dashboard—a quarterly review of these four metrics is sufficient for most systems.

1. **Extraction or processing accuracy:** sample a share of outputs against source data each month.
2. **Exception rates:** a sudden spike, or a drop to zero, both signal the system may have changed behavior.
3. **Processing volume:** anomalies can flag a data-feed issue or system malfunction.
4. **Time from a vendor update to re-verification:** the gap most banks miss, and the one that would have caught the loan-document scenario above.

> **Key Governance Actions for Back-Office AI:**
> Include every back-office AI system in the inventory—even those your vendor labels as "automation" rather than "AI." Assign an owner. Establish a sample-based output verification process proportionate to the system's risk tier. Require your vendor to notify you before any update that changes how the system processes data. And add a post-update verification step to your change management process so that vendor updates do not silently break AI outputs that feed into downstream decisions.

15.5 Generative AI for Employee Productivity

Generative AI for internal productivity—drafting communications, summarizing documents, researching topics, generating code, analyzing data—is the fastest-growing AI use case in community banking and the one most likely to be adopted informally without governance. Chapter 7's generative AI policy and Chapter 10's enterprise management guidance provide the structure. This section addresses the governance considerations specific to common productivity use cases.

DRAFTING AND COMMUNICATION

Employees using generative AI to draft emails, memos, reports, or customer communications should be required to review all output for accuracy, tone, and appropriateness before use. For customer communications, the reviewer should confirm that all facts are accurate, disclosures are complete, and the communication complies with applicable regulations. The rule that covers every case: nothing AI-drafted leaves the bank without a person having owned it first.

RESEARCH AND ANALYSIS

Generative AI can be a powerful research tool, but its outputs can include fabricated facts, invented citations, and plausible-sounding but incorrect analysis. Employees using generative AI for research should be trained to verify all facts and citations independently, treat AI-generated analysis as a starting point rather than a conclusion, and never rely solely on AI-generated research for regulatory submissions, customer recommendations, or strategic decisions.

DATA ANALYSIS

Using generative AI to analyze bank data—financial reports, customer data, portfolio metrics—raises particular governance concerns. The data input restrictions in the acceptable use policy must be enforced rigorously. Employees should use only the bank's approved enterprise AI platform for data analysis, never consumer tools. Analysis results should be validated against source data before being used for decision-making.

AI CODING ASSISTANTS

IT teams in community banks are increasingly using AI coding assistants—tools like GitHub Copilot, Amazon Q Developer, or similar platforms—to accelerate software development, automation scripting, and system integration work. These tools present specific governance considerations beyond the general generative AI policy.

- **Code security:** AI-generated code can contain security vulnerabilities, hardcoded credentials, or insecure patterns learned from the tool's training data. Every piece of AI-generated code deployed in the bank's production environment must undergo the same security review applied to human-written code. For code that touches financial transactions, customer data, or system administration functions, enhanced review is warranted.

- **Intellectual property:** AI coding assistants are trained on publicly available code repositories, some of which contain code with restrictive licenses. Code generated by these tools may inadvertently incorporate licensed material. The IT team should understand the licensing implications of the AI coding tool they use and implement review procedures to identify potential IP issues before deploying generated code.

- **Dependency and maintainability:** AI-generated code that no human on the team fully understands creates a maintainability risk. When something breaks, someone needs to debug it. The governance policy should require that any AI-generated code deployed in production is reviewed by a qualified human who understands what the code does and can maintain it going forward. AI coding assistants should accelerate development without creating technical debt that the team cannot service.

AGENTIC AI WORKFLOWS

Agentic AI—a system that plans and executes multi-step tasks autonomously—is beginning to appear in the productivity context: drafting-and-sending workflows, scheduling, document routing, and IT administration tasks. Chapter 11 sets out the full governance model—bright lines, identity and access rules, autonomy caps, and the kill switch. In the productivity context, two of its controls matter most: an approval gate before anything leaves the bank (an email sent, a document filed, a calendar invitation issued), and complete action logs so a reviewer can reconstruct what the agent did. Treat any agentic feature that appears in a productivity platform as a new system for Chapter 8 inventory purposes, even when it arrives inside a tool the bank has already approved.

Looking Ahead on Agentic AI

Agentic AI is arriving through the platforms you already run—vendors are embedding agentic capabilities into productivity suites and core systems, and your IT team may already be experimenting with agents for administration tasks. The time to adopt Chapter 11's controls is before the first agent goes live—they are far easier to design in than to retrofit. Chapter 17's discussion of emerging technology trends provides the broader strategic context.

15.6 AI in Marketing and Customer Analytics

AI-powered marketing—segmentation, personalized offers, propensity modeling, targeted advertising—earns its keep by putting the right product in front of the right customer. The same engine that targets can exclude, which is why these applications run into fair lending, privacy, and consumer protection faster than most banks expect.

FAIR LENDING CONSIDERATIONS

AI-driven marketing that targets certain customer segments for product offers—or excludes certain segments from offers—can raise fair lending concerns even before a credit decision is made. If the bank's marketing AI systematically excludes certain geographic areas or demographic groups from seeing offers for credit products, the result is a form of digital redlining that can trigger regulatory scrutiny. The governance program should require periodic analysis of marketing AI outcomes to ensure that offer distribution is equitable across protected classes.

PRIVACY AND CONSENT

Marketing AI often relies on detailed customer data—transaction history, account balances, browsing behavior, location data—to generate personalized offers. The bank must ensure that its use of customer data for AI-driven marketing is consistent with its privacy notices and compliant with applicable law. Customers should have the ability to opt out of AI-driven marketing personalization, and their opt-out preferences must be respected across all channels.

GOVERNANCE REQUIREMENTS

Marketing AI answers to five reviews. Governance should include periodic fair lending analysis of marketing outcomes across protected classes, privacy compliance review ensuring data use aligns with published privacy notices and customer consent, content review of AI-generated marketing messages for accuracy and regulatory compliance, opt-out mechanism testing to confirm that customer preferences are being honored, and documentation of marketing AI targeting logic and any exclusion criteria.

> **Integration Reminder**
> Marketing AI rides rails the bank already has: the same compliance review that clears a traditional campaign clears an AI-driven one, with three additions—algorithmic targeting, bias, and personalization. A separate AI marketing process would be one more thing to maintain and one more place to miss.

15.7 AI in Compliance and Regulatory Monitoring

Compliance is where community banks feel the squeeze most—the rulebook keeps growing and the team does not. AI is increasingly taking on compliance work that once required manual review, and for a lean team it is one of the highest-value use cases available. The governance challenge is that compliance AI must be held to the highest accuracy standard—an error in compliance monitoring becomes a regulatory finding.

How AI Is Being Used

- Automated control testing is one of the most immediate applications. AI tools can continuously monitor whether internal controls are operating as designed—testing transaction limits, verifying segregation of duties, and flagging exceptions that would previously surface only during a quarterly review or an annual audit. For a community bank with a two-person compliance team, this shifts the posture from periodic review to continuous monitoring without adding head count.

- Regulatory change management is another emerging use case. AI systems can monitor federal and state regulatory publications, identify changes relevant to the bank's operations, and flag provisions that may require policy updates or operational adjustments. The pace of state-level AI legislation makes this capability increasingly valuable: Colorado's 2026 rewrite of its AI Act (detailed in Section 17.2) is the leading example; Texas, Illinois, and others are advancing their own requirements, and federal preemption remains under debate in Congress—leaving the state patchwork as the practical regulatory reality. For community banks operating across state lines, automated regulatory monitoring has become a necessity.

- Examination preparation is a third area where AI adds value. AI tools can organize and pre-assemble documentation that examiners typically request, identify gaps in the bank's documentation before examiners find them, and generate preliminary self-assessments that help management anticipate examination focus areas.

Governance Considerations

AI in compliance carries a specific governance risk: over-reliance. If the compliance team begins trusting AI outputs without independent verification, the bank has effectively delegated a regulatory obligation to an algorithm. The governance program must ensure that AI-generated compliance findings are reviewed by qualified compliance staff before action is taken, that the AI system's logic is documented and auditable, and that the bank can demonstrate to examiners that human judgment remains central to compliance decisions. AI should augment the compliance function while the judgment—and the accountability—stay with people.

15.8 AI-Enhanced Cybersecurity

AI is reshaping cybersecurity in two directions simultaneously: it is the most powerful defensive tool community banks have access to, and it is the most dangerous offensive weapon being used against them. According to the CSI 2026 Banking Priorities report, AI-enhanced social engineering attacks—including voice cloning, deepfake video, and AI-generated phishing—jumped sixteen percentage points to become the leading cybersecurity concern for financial institutions in 2026. Community banks cannot afford to think about AI cybersecurity as a future problem. It is the current threat landscape.

AI as a Defensive Tool

AI-powered security systems can analyze network traffic patterns, detect anomalies in real time, identify suspicious login behavior, and correlate threat indicators across multiple data sources faster than any human team the bank could hire. For community banks with small or outsourced IT security teams, AI-driven threat detection can provide a level of continuous monitoring that would otherwise require a dedicated security operations center. Many community banks are already receiving these capabilities through their managed security service providers, endpoint detection platforms, and email filtering systems—often without realizing that AI is the engine behind them.

AI as an Offensive Threat

The same AI capabilities that power defensive tools are being weaponized by attackers. Voice cloning technology can replicate a CEO's voice from a few seconds of publicly available audio, enabling convincing vishing attacks that bypass traditional social engineering defenses. AI-generated phishing emails are increasingly sophisticated—grammatically perfect, contextually relevant, and personalized to the recipient using scraped social media data. Deepfake video is emerging as a tool for business email compromise, where an attacker impersonates a trusted executive in a video call to authorize a fraudulent wire transfer.

Governance Considerations

AI cybersecurity governance has to cover both directions.

On the defensive side, the bank must:

- inventory and assess the AI components within its security tools;
- ensure AI-driven alerts are reviewed by qualified staff before they are trusted;
- validate that the AI models are tuned for the bank's specific environment.

On the offensive side, the bank must:

- require out-of-band verification before any request to move money or reset credentials is honored, so a cloned voice cannot authorize action by itself;
- harden email authentication and keep staff trained against AI-generated phishing, so a grammatically perfect, personalized message still draws scrutiny;
- apply dual control to wire transfers and executive payment requests, so a deepfake video call cannot stand in for authorization.

Section 15.9 takes up the full attack landscape and the verification controls that counter it.

15.9 AI-Enabled Fraud: Deepfakes, Voice Cloning, and Synthetic Identity

Sections 15.2 and 15.8 examined AI as a tool the bank uses to detect fraud and defend its systems. This section turns the lens around: the fraud that AI now enables against the bank and its customers, and the controls that hold up when a voice, a face, or a document can no longer be trusted as proof of identity.

THE SCALE OF AI-ENABLED FRAUD

The numbers leave no room for complacency. The FBI's Internet Crime Complaint Center reported a record $20.9 billion in losses for 2025—a 26 percent jump over the prior year and the first time the figure has crossed $20 billion in the center's twenty-five-year history. Investment fraud ($8.6 billion) and business email compromise ($3.0 billion)—the two categories most directly amplified by AI—led the totals. For the first time, the report carried a dedicated section on AI-enabled crime, logging 22,364 complaints and $893 million in losses in its debut year of tracking—a baseline that will almost certainly look small in hindsight.

Looking forward, Deloitte's Center for Financial Services has projected that generative AI could push U.S. fraud losses to roughly $40 billion by 2027, up from an estimated $12.3 billion in 2023. Whether or not that specific projection holds, the direction is unmistakable: AI lowers the cost, raises the quality, and multiplies the scale of fraud aimed at banks and their customers.

For a community bank, the uncomfortable truth is that smaller institutions are often easier to target, and attackers know it. Fewer verification layers, more informal approval processes, and closer personal relationships (which attackers can study and imitate) are exactly the conditions AI-enabled fraud is designed to exploit.

SYNTHETIC IDENTITY FRAUD

Synthetic identity fraud—described by the Federal Reserve as one of the fastest-growing financial crimes in the United States—combines real and fabricated information (often a real Social Security number, frequently a child's or an elderly person's, paired with a fictitious name and date of birth) to create a "person" who does not exist. AI accelerates every stage: generating convincing application data, fabricating supporting documents, and even producing synthetic faces and IDs that defeat automated identity verification. Because the identity is not stolen from a real victim who will complain, synthetic identities can sit quietly in a portfolio for months or years—building credit, then "busting out" with maxed balances and disappearing.

For community banks, the governance implications fall squarely on account-opening and lending controls: identity-verification vendors should be assessed for their ability to detect synthetic identities and AI-generated documents, and the AI inventory should flag any onboarding tool that relies on automated document or biometric verification as a system whose failure mode is now adversarial.

AUTHORIZED PUSH PAYMENT (APP) FRAUD AND THE LIABILITY GAP

In a growing share of fraud, the customer is not hacked—they are deceived into authorizing the payment themselves. AI-generated romance scams, investment scams, and impersonation scams persuade customers to send money through wires, Zelle, or other faster-payment rails. This "authorized push payment" fraud sits in a regulatory gap: under Regulation E, banks generally must reimburse unauthorized electronic transfers, but a payment the customer was tricked into authorizing is, on its face, authorized—and typically not reimbursable. The federal posture remains unsettled: the CFPB's December 2024 suit against the operator of Zelle and three large banks over induced fraud was voluntarily dismissed in March 2025. The United Kingdom, by contrast, adopted a mandatory APP-reimbursement regime in October 2024. Community banks should not wait for the U.S. rules to settle: customer-facing warnings at the point of payment, transaction-hold and confirmation steps for high-risk transfers, and front-line training to spot scam patterns are the practical defenses—and they double as evidence of a bank acting in good faith.

ELDER FINANCIAL EXPLOITATION

Older customers are both disproportionately targeted and disproportionately harmed. In an April 2024 Financial Trend Analysis covering a single year of Bank Secrecy Act filings, FinCEN identified roughly $27 billion in suspicious activity tied to elder financial exploitation, and AI sharpens the oldest tricks: a voice-cloned "grandchild" in distress, a deepfaked "government official," or a synthetic romance built over months. Community banks—where a teller often knows the customer personally—hold a real advantage here. Front-line staff trained to recognize the behavioral signs of exploitation (urgency, secrecy, unusual withdrawals, a new "advisor" on the phone), paired with transaction holds and trusted-contact procedures, are frequently the last and best line of defense. This is governance that aligns perfectly with the community banking mission.

DEFENSIVE VERIFICATION: ASSUME THE VOICE, FACE, OR DOCUMENT MAY BE SYNTHETIC

The same technology powering these attacks also powers the defense. AI-driven transaction monitoring, behavioral biometrics, and anomaly detection can flag the subtle patterns of synthetic-identity and scam activity faster than rules-based systems. But the decisive control is procedural: verification steps that hold up even when the voice, face, or document may be synthetic. They include:

- **Callback verification:** use a pre-established phone number, never one supplied in the suspicious message.
- **Out-of-band confirmation:** for wire and high-value payment changes.
- **Multi-person approval:** for large transfers.

- **An operating assumption:** that a convincing voice or video is no proof of identity.

FinCEN's 2024 alert on deepfake-enabled fraud (FIN-2024-Alert004) catalogs red flags worth building into both training and monitoring—inconsistencies between a customer's identity documents and other information, and reverse-image-search "hits" suggesting AI-generated photos among them—and supplies the SAR keyword (FIN-2024-DEEPFAKEFRAUD) for reporting suspected incidents.

Chapter Summary

This chapter has applied the governance program to the eight AI use cases most common in community banking—and to the AI-enabled fraud aimed at them:

- Lending AI demands the highest governance rigor: comprehensive documentation, independent validation, fair lending testing, compliant adverse action notices, and meaningful human review of denials and borderline decisions.
- Fraud detection and BSA/AML AI require documented tuning decisions, effectiveness reviews, false positive management, and human SAR decision-making.
- Customer-facing AI needs disclosure, output monitoring, escalation rules, and regular content review.
- Back-office AI receives proportionally lighter governance focused on inventory, ownership, and output verification.
- Generative AI productivity tools require acceptable use enforcement, human review of all outputs, data input restrictions, and employee training.
- Marketing AI must be governed for fair lending, privacy compliance, content accuracy, and opt-out effectiveness.
- Compliance AI can automate control testing, regulatory monitoring, and examination preparation—but must be governed against over-reliance that effectively delegates regulatory obligations to an algorithm.
- AI cybersecurity operates on both sides of the equation: as a defensive tool providing continuous threat monitoring and as an offensive weapon—voice cloning, deepfake, AI-generated phishing—that community banks must train and build controls against.

- AI-enabled fraud—synthetic identity, deepfakes, voice cloning, and authorized push payment scams—demands verification procedures that assume a voice, face, or document may be synthetic, transaction holds and front-line training, and particular vigilance for elder exploitation.

Looking Ahead

The remainder of Part VI turns to sustaining the program over time. Chapter 16 addresses the human side—training, change management, and the culture that determines whether governance is practiced or merely documented. Chapter 17 then looks ahead to the technologies and regulatory changes the program must be ready to absorb.

Evaluate Before You Deploy

Every use case in this chapter carries a different risk profile. The AI Governance Toolkit includes the AI Use-Case Intake Form and Evaluation Scorecard—an Excel workbook for weighing business value against regulatory risk—so your governance committee can make informed deployment decisions.

"You can write a perfect policy, build an elegant governance structure, and design a flawless risk assessment process. If nobody follows them, all you have is a filing cabinet."

— COMMUNITY BANK RISK OFFICER —

✦ CHAPTER 16 ✦

Building an AI-Aware Culture

PEOPLE MAKE GOVERNANCE WORK

Governance structures exist on paper. Culture exists in behavior. The most comprehensive AI governance program in the world will fail if the people it depends on—employees, managers, committee members, board directors—do not understand it, believe in it, and follow it in their daily work. This chapter addresses the human side of AI governance: training, change management, innovation culture, and talent strategy.

16.1 Training Programs by Role

Effective AI training is fitted to the job: a board member, a loan officer, and an IT specialist each need different knowledge, at different depths, delivered in different formats. The following training matrix can be adapted to each bank's organizational structure.

AUDIENCE	CORE TOPICS	DEPTH	FREQUENCY
Board of Directors	AI landscape, governance program, risk posture, fiduciary obligations	Strategic awareness; no technical detail required	Annual briefing + quarterly report review
Senior Management	All board topics + governance operations, policy implementation, incident management	Working knowledge sufficient to manage and challenge	Semi-annual training + ongoing committee participation
IT/Technical Staff	AI technology, security threats, monitoring, validation, vendor management	Technical proficiency appropriate to role	Quarterly updates + specialized training as needed
Compliance / Risk	Regulatory requirements, fair lending testing, model risk management, policy compliance	Working knowledge of AI risk and regulatory expectations	Semi-annual training + regulatory update briefings
Front-Line Staff	Acceptable use policy, approved tools, data handling, output verification, reporting obligations	Practical knowledge to use AI responsibly in daily work	Annual training + refreshers when policies change

Three principles should guide the design of AI training across all audiences.

1. **Relevance:** every training session should connect AI governance to the participant's actual job responsibilities. Abstract lectures on machine learning theory will not change behavior. Concrete

scenarios—what to do when a customer asks about the chatbot's advice, how to handle a vendor's AI upgrade notification, when to escalate an anomalous model output—will.

2. **Accessibility:** training should use plain language, avoid unnecessary jargon, and acknowledge that most participants are encountering these concepts for the first time. The goal is to build the minimum viable understanding each role requires.

3. **Reinforcement:** a single training session creates awareness; sustained reinforcement creates culture. Follow initial training with periodic refreshers, policy reminders embedded in regular communications, practical exercises, and visible leadership engagement with AI governance topics.

FRONT-LINE TRAINING: THE CRITICAL PRIORITY

Front-line employees—tellers, loan officers, customer service representatives, operations staff—are the largest employee group and often the most exposed to AI governance risks. They are the ones who might paste customer data into ChatGPT to draft a letter, or accept a vendor's AI credit recommendation without meaningful review, or fail to disclose to a customer that they are interacting with a chatbot. Training for this group should be practical, scenario-based, and refreshed annually at minimum.

Consider building training around real scenarios: a loan officer receives an AI-generated credit denial—walk through how to review the recommendation, what questions to ask, when to override, and how to explain the decision to the customer. A customer service representative discovers a colleague using an unapproved AI tool—walk through the reporting process and emphasize the no-blame reporting culture. An operations specialist receives a vendor notification about an AI upgrade—walk through the steps to escalate to IT for governance review.

Measuring Training Effectiveness

Training completion rates tell you who attended; who learned is a separate question. Supplement completion tracking with periodic knowledge assessments (brief quizzes or scenario-based tests), behavioral indicators (are employees actually using approved tools and following data handling procedures?), and incident analysis (are incidents declining, and are employees reporting concerns proactively?). Section 16.5 extends the same logic to culture itself.

A first-year rollout can be modest and still complete. One workable sequence:

WHEN	AUDIENCE	FOCUS
Month 1	All staff	AI fundamentals; acceptable use policy attestation
Month 2	Lending and front-line teams	Scenario training: AI outputs, overrides, and documentation
Month 3	Governance committee	Risk methodology deep dive (Chapter 8) and dashboard walkthrough
Quarter 2	Board of directors	Annual AI education session (Section 13.5)
Quarter 3	Operations and back office	Data handling, shadow AI, and reporting channels
Quarter 4	All staff	Refresher with incident lessons; attestation renewal

The Toolkit's AI Training Rollout Plan provides the full version of this calendar with session outlines and the attestation form.

16.2 Change Management for AI Adoption

AI adoption is a change management challenge as much as a technology challenge. Employees may resist AI for understandable reasons: fear that AI will replace their jobs, discomfort with unfamiliar technology, skepticism about AI's reliability, or resentment of what feels like additional bureaucratic burden. Effective change management addresses these concerns honestly and proactively.

ACKNOWLEDGING CONCERNS

The worst thing leadership can do is dismiss employee concerns about AI. Fear of job displacement is legitimate, even if the bank has no intention of reducing staff. Skepticism about AI accuracy is warranted—AI systems do make mistakes. Frustration with new policies and procedures is natural when they add steps to established workflows. Leadership should acknowledge these concerns openly, provide honest information about the bank's AI strategy and its workforce implications, and create channels for employees to ask questions and voice concerns.

Here is what this sounds like in practice. A senior loan officer says: "I've been making credit decisions for twenty years. I don't need a machine second-guessing me, and I don't have time to fill out a new form every time the system gives me a score." This is a professional expressing legitimate concern about autonomy and workload. The effective response addresses the objection directly: the AI gives you an additional data point and leaves the judgment yours; the governance documentation protects you personally if a decision is ever questioned; and the form takes three minutes. When you address the real concern behind the resistance, compliance follows. When you dismiss it, you get shadow workarounds.

BUILDING CHAMPIONS

Successful change management leverages early adopters and influencers within the organization. Identify employees who are curious about AI, provide them with early access to approved tools and training, and empower them to help their colleagues. These champions become peer educators, first-line troubleshooters, and proof at the next desk over that the approved tools are worth using. In community banks, where personal relationships drive organizational dynamics, peer influence is often more powerful than top-down mandates.

COMMUNICATING THE WHY

Employees are far more likely to embrace AI governance if they understand why it matters—not just to the bank's compliance posture but to them personally and to the customers they serve. Frame AI governance as the mechanism that protects the bank's reputation, protects customers from unfair or inaccurate decisions, protects employees from inadvertently creating risk, and enables the bank to adopt helpful technology confidently. When employees see governance as protective rather than restrictive, adoption follows naturally.

MAKING IT SAFE TO REPORT

More than one of this book's case studies turns on a single variable: whether an employee speaks up when something feels wrong. That variable has to be designed. A no-blame reporting channel has three required features: a known intake point (a named person and a simple form), a guaranteed response (written acknowledgment and a one-line outcome note), and a visible separation between reporting and discipline. The Ridgeview loan officer in Case Study 7 was retrained rather than terminated; that outcome, repeated and made known, is the real policy.

The boundary needs stating too. Protection covers good-faith reports and self-reports, including embarrassing ones; concealment, repeated knowing violations, and misuse for harassment sit outside it. Publish both halves and the channel will be used. Publish only the discipline half and Chapter 2's shadow-AI footprint grows in silence.

16.3 Encouraging Responsible Innovation

Governance must not become a synonym for "no." If employees perceive the governance program as a barrier to innovation, they will work around it—creating exactly the shadow AI problem the program is designed to prevent. The governance program should include mechanisms for encouraging and channeling innovation within appropriate boundaries.

THE INNOVATION INTAKE PROCESS

Create a simple, accessible process for employees to propose AI use cases. The process should be lightweight enough that employees are not deterred from submitting ideas but structured enough to capture the information the governance committee needs for initial evaluation. A brief intake form—business problem, proposed AI solution, expected benefit, data involved, and perceived risks—is sufficient. The governance committee or a designated reviewer should respond to every submission within a defined time frame, even if the response is "we're adding this to the evaluation queue."

SANDBOXED EXPERIMENTATION

For banks with the technical capacity, providing a sandboxed environment where employees can experiment with approved AI tools using non-sensitive data allows exploration without risk. The sandbox should have clear boundaries: no customer data, no connection to production systems, and no deployment without governance review. But within those boundaries, employees should be free to explore, learn, and develop potential use cases that can be brought to the governance committee for formal evaluation.

RECOGNIZING AND REWARDING RESPONSIBLE USE

Culture is shaped by what the organization celebrates. Recognize employees who identify productive AI use cases, who report governance concerns proactively, who help colleagues use approved tools effectively, and who demonstrate the values of responsible AI use. This does not require elaborate reward programs—public acknowledgment, a mention in a team meeting, or a note from senior management is usually enough. People repeat what gets noticed.

> **The Innovation-Governance Balance**
>
> Zero risk would require zero innovation. The goal is informed risk-taking within a governance program that ensures risks are identified, assessed, and managed before they become problems. The governance committee that approves a well-assessed moderate-risk AI use case is doing its job just as much as the committee that rejects a poorly assessed high-risk proposal.

16.4 Talent Strategy for AI Governance

Sustaining effective AI governance requires skills that many community banks do not yet possess in-house. The talent strategy should address this gap through a combination of internal development, external partnerships, and strategic hiring.

DEVELOPING INTERNAL TALENT

Most AI governance skills build on existing competencies. A compliance officer who understands model risk management can learn AI-specific risk assessment. An IT manager who understands information security can learn AI-specific threat vectors. A risk analyst who understands statistical testing can learn bias detection methodology. The most cost-effective strategy is developing the staff you already trust—the cheapest expert is the one you already employ, six months of focused development from now.

USING EXTERNAL EXPERTISE

Community banks do not need to hire full-time AI specialists to maintain effective governance. The bank can rent what it does not need to own. Consulting firms now specialize in AI governance for community banks. Model validation can go to a qualified outside shop. Industry associations pool shared resources and peer learning, the regulators themselves publish guidance and give examination feedback, and technology vendors will train your people on the AI products they sold you.

The key is to use external expertise strategically, supplementing capabilities the bank builds and keeps in-house. The bank must maintain sufficient internal understanding to manage its external advisors effectively, interpret their findings, and make governance decisions. The work can be outsourced; the judgment must stay in the bank.

LONG-TERM HIRING CONSIDERATIONS

As AI use grows, some community banks will find it necessary to add staff with AI-specific expertise. This might be a data analyst who can support model monitoring and bias testing, an IT specialist with AI security knowledge, or a risk professional with model risk management experience. When hiring, prioritize candidates who combine technical knowledge with the ability to communicate effectively with non-technical colleagues—the translation skill governance runs on.

MAKING THE BUILD-VS.-BUY DECISION

Most community banks will not justify a full-time AI governance hire in the near term. The economics favor a blended approach:

- **Build internal AI literacy across existing roles.**

- **Outsource specialized tasks—model validation, bias testing—to qualified third parties.**

- **Revisit the hiring decision annually as your AI portfolio grows.**

A practical benchmark is this—if your bank has five or more High- or Critical-tier AI systems, the governance workload will likely justify a dedicated resource or a meaningful reallocation of an existing role. Below that threshold, a well-trained IT leader or compliance officer who dedicates a portion of their time to AI governance, supported by external specialists for validation and testing, is both cost-effective and defensible to examiners. The key is to document the decision and the rationale, so that your governance program shows deliberate resource allocation rather than benign neglect.

16.5 Measuring Culture

This chapter has argued that culture determines whether governance is practiced. That claim is testable. Four signals, tracked quarterly, tell the committee whether the culture is taking hold:

1. **Policy acknowledgment and training currency—the share of staff current on attestations and role-based training. A reasonable target is above 95 percent, refreshed annually.**

2. **Innovation intake volume—use-case proposals submitted through the Section 16.3 process. A healthy pipeline at a community bank is a handful per quarter; zero means employees are either unaware of the channel or using AI without telling you.**

3. **Proactive incident and concern reports—employees flagging AI issues before monitoring catches them. Rising reports in the first year are a good sign: the channel is trusted. The worrying number is zero.**

4. **Override and escalation documentation—whether staff record their reasoning when they override model outputs, sampled**

from Section 8.4's monitoring data. Documentation rates show whether governance has reached daily work.

One illustration of why the third signal matters: a bank's quarterly review showed proactive incident reports falling from six to zero over two quarters. Leadership initially read the decline as success. The training survey told a different story—staff had concluded that reporting a near-miss invited paperwork with no visible response. The fix was procedural and small—the guaranteed-response feature described in Section 16.2, installed after the fact: every report now receives a written acknowledgment and a one-line outcome note. Reports resumed, and two of the next quarter's reports caught real configuration drift. The metric was measuring trust, and trust is recoverable.

Report the four signals quarterly on the Chapter 12 dashboard, and revisit targets annually as part of the Section 12.6 governance review.

Chapter Summary

This chapter has addressed the human dimensions of AI governance—the training, change management, and talent strategies that determine whether governance takes hold:

- Role-based training—from the board to front-line staff—builds the minimum viable understanding each position requires, with practical scenarios that connect governance to daily responsibilities.
- Change management addresses employee concerns honestly, builds internal champions, and frames governance as protective rather than restrictive.
- Responsible innovation channels—intake processes, sandboxed experimentation, and recognition—prevent governance from becoming a barrier to productive AI adoption.
- A talent strategy combining internal development, external partnerships, and strategic hiring ensures the bank has the skills to sustain governance over time.
- Culture is measurable: policy acknowledgment and training currency, innovation intake volume, proactive incident reports, and override documentation—the four signals of Section 16.5.

Looking Ahead

Chapter 17 concludes the book by looking forward—emerging technology trends, the evolving regulatory frontier, industry collaboration, and a governance maturity roadmap for the years ahead.

Culture Change Needs a Plan

The AI Governance Toolkit includes the AI Training Rollout Plan—with an employee attestation form and a ready-to-deliver awareness training deck—so the training program ships the same week the policy does.

"The best time to build a governance program was before you started using AI. The second best time is now. The worst time is after the examiner asks for it."

— COMMUNITY BANK PRESIDENT —

✦ CHAPTER 17 ✦

Future-Proofing Your AI Governance

PREPARING FOR WHAT'S NEXT

This book has provided a comprehensive program for AI governance in community banking—from foundational principles through policies, risk assessment, vendor management, implementation playbooks, ethics, audit, use cases, and culture. This final chapter looks forward, addressing the emerging technologies that will expand the governance challenge, the regulatory developments that will reshape the compliance landscape, the industry collaboration that can accelerate governance maturity, and a multi-year roadmap for maturing governance from Foundational to Advanced.

The chapter closes with a reflection on why AI governance is, at bottom, an expression of the values that have always defined community banking.

17.1 Emerging Technology Trends

The AI landscape is evolving at a pace that makes specific predictions unreliable. Rather than forecasting which technologies will dominate, this section identifies the trends most likely to affect community banking and the governance implications of each.

AGENTIC AI

Agentic AI systems can autonomously plan and execute multi-step tasks—researching, analyzing, deciding, and acting with minimal human intervention. In banking, agentic capability is moving from pilot to platform: vendors are shipping it inside core, compliance, and productivity systems on their own release schedules. Chapter 11 established the governance model—bright-line decisions, agent identity and access rules, autonomy caps, contract provisions, and monitoring metrics. The trend to watch from here is scope: how quickly vendor platforms expand what agents are permitted to do, and whether each expansion arrives with the disclosure and opt-in controls Section 11.4 calls for. Expect examiner interest to follow the same path, with early attention to kill-switch discipline and the audit trail behind autonomous actions.

MULTIMODAL AI

Current AI systems primarily process text and structured data. Multimodal AI processes multiple data types simultaneously—text, images, audio, and video. In a community bank, that looks like an underwriter whose system reads the property photos alongside the financials, a service platform that follows one customer from a phone call into a chat window, and software that pulls usable numbers off a handwritten receipt. The governance program should be prepared to assess multimodal AI systems that consume richer, more complex data—with correspondingly more complex privacy, accuracy, and bias considerations.

EMBEDDED AND INVISIBLE AI

The trend toward embedding AI deeply into software platforms will accelerate. AI features will become standard components of every banking technology platform, activated by default and evolving continuously through vendor updates. For governance, this means the inventory challenge described in Chapter 8 will become more complex. Banks will need systematic processes for identifying new AI capabilities as they emerge in existing vendor platforms and assessing their governance implications promptly.

AI-GENERATED SYNTHETIC DATA

AI systems that generate synthetic data—artificial datasets that mimic the statistical properties of real data without containing actual customer information—promise a way around three hard problems: privacy, model testing, and bias mitigation. They also bring questions of their own: Does synthetic data accurately represent the real customer base? Could synthetic data introduce biases not present in real data? How should synthetic data be governed under data governance and model risk management frameworks? When a vendor proposes synthetic data—for testing, for privacy, for bias mitigation—run it through the same risk assessment as real data, with those three questions added.

> **Governance Adaptability**
> Technologies that do not yet exist need no governance today. What you need is a program flexible enough to accommodate them when they arrive. The principles-based approach in this book—proportionality, transparency, fairness, accountability, privacy, and safety—provides that flexibility. New technologies create new risks, but the questions remain the same: What does this system do? What are the risks? How are they managed? Who is accountable?

17.2 The Evolving Regulatory Frontier

The regulatory landscape for AI in banking will change significantly in the coming years. While the specific direction of regulation is uncertain, several trends are clear enough to inform governance planning.

INCREASED SUPERVISORY FOCUS

Federal banking agencies have signaled unambiguously that AI is a supervisory priority. Examination procedures will become more specific and more rigorous as regulators develop AI-specific expertise. Community banks that have proactively built governance programs will be well positioned; those that have delayed will face increasing pressure. This is no longer a prediction—the April 2026 release of SR 26-2, which replaced SR 11-7 with a materiality-based, risk-proportionate model risk framework, is a concrete example of supervisory expectations becoming more specific to how banks actually use AI.

MORE SPECIFIC REGULATORY GUIDANCE

The current regulatory framework relies heavily on adapting existing guidance—model risk management, fair lending, vendor management—to AI contexts. Regulators are likely to issue more specific AI guidance that addresses gaps in existing frameworks, such as governance expectations for generative AI, standards for AI explainability in consumer decisions, and requirements for AI bias testing methodology. The annual policy review on the committee's calendar (Section 12.1) is the natural absorption point: new guidance lands, the program takes it in, and the bank moves on. The Treasury's Financial Services AI Risk Management Framework, released in February 2026 with approximately 230 control objectives tailored to financial institutions, is an early example of this trend—and community banks should expect similar sector-specific frameworks from the banking agencies themselves in the near term.

STATE-LEVEL ACTIVITY

State legislatures are moving faster than federal agencies on AI regulation. State laws addressing algorithmic transparency, automated decision-making, and AI bias are proliferating, and some of these laws may apply to banking activities not preempted by federal law. Community banks that operate in multiple states or that use AI systems that touch state-regulated activities should know which state regimes reach them—the annual scope check from Section 4.5 covers it. Whether Congress ultimately preempts state AI regulation remains an open question as this book goes to press; until it acts, the state patchwork is the practical reality.

Colorado's SB 26-189, signed in May 2026, replaced the state's landmark 2024 AI Act with a narrower regime—effective January 1, 2027—requiring notice when consumers interact with AI, disclosure after adverse automated decisions, data-correction rights, and meaningful human review, while dropping the earlier law's impact-assessment and risk-program mandates. Meanwhile, state attorneys general are actively enforcing existing consumer protection and fair lending statutes against AI-driven decisions—Massachusetts's $2.5 million settlement with a fintech lender over algorithmic lending discrimination in July 2025 demonstrated that states will act where they see harm, regardless of the federal enforcement posture.

PREPARING FOR REGULATORY CHANGE

The most effective preparation for regulatory change is a governance program built on sound principles and flexible processes. A bank with a comprehensive AI inventory, tiered risk assessments, documented policies, and established monitoring can incorporate new regulatory requirements incrementally. A bank starting from scratch when a new rule is finalized faces a far more expensive and disruptive implementation. Governance built today is regulatory cost avoided tomorrow.

17.3 Industry Collaboration and Peer Learning

Community banks do not need to build AI governance in isolation. An association roundtable, a peer bank that solved your problem last year, a shared template—each one shortens the timeline and splits the cost of building governance.

INDUSTRY ASSOCIATIONS

Organizations like the Independent Community Bankers of America, state banking associations, and other industry groups are developing AI governance resources tailored to community banks. The output that matters is concrete: model policies, training curricula, peer roundtables, and regulatory advocacy. One state association's quarterly AI roundtable, to take a representative example, produced a shared vendor-question library and a model AI policy that member banks adapted in weeks instead of writing from scratch. Active participation buys the bank resources it could never afford to develop alone—and a bench of peers who have already met the problem you are about to.

PEER NETWORKS

Community banks in the same region or of similar size often face identical AI governance challenges. Informal peer networks—whether organized through industry associations, regulatory-hosted roundtables, or direct relationships—provide a valuable forum for sharing experiences, comparing approaches, and learning from each other's successes and mistakes. These conversations are often more practical and relevant than formal conferences because they involve institutions with similar resource constraints and risk profiles.

SHARED SERVICES AND RESOURCES

Some community banks are exploring shared approaches to AI governance activities that are expensive to perform independently, such as model validation, bias testing, and vendor AI due diligence. Consortia or cooperative arrangements pool resources to engage specialized expertise, develop shared assessment frameworks, and cut per-bank costs. Shared services demand care on confidentiality and independence, and they put capabilities like independent model validation within reach of institutions that could never fund them alone.

17.4 Building a Governance Maturity Roadmap

AI governance maturity develops over time. No community bank will implement every element of this book simultaneously, and no bank should try. The following maturity roadmap provides a realistic, multi-year path from initial governance to advanced practice.

STAGE	TIMELINE	KEY MILESTONES	SELF-ASSESSMENT INDICATORS
Initial (starting point)	Present Day	None yet—ad hoc AI use without inventory, policy, or oversight. The 90-Day Quick Start is the path out.	No stage indicators met; scored 0—Initial in the companion Toolkit's assessment.
Foundational	Months 1–6	AI inventory, acceptable use policy, governance committee formed, initial risk assessments	Inventory exists; policies approved by board; committee meeting regularly

STAGE	TIMELINE	KEY MILESTONES	SELF-ASSESSMENT INDICATORS
Evolving	Months 6–12	Full policy suite, tiered risk system operational, vendor due diligence program, board reporting established	All policies in place; assessments completed for High/Critical systems; quarterly board reports
Established	Year 2	Monitoring and validation programs mature, bias testing routine, incident response tested, audit coverage initiated	KPIs tracked for all High/Critical systems; risk-based validation reviews completed on schedule; tabletop exercise completed; audit findings addressed
Advanced	Year 3+	Continuous improvement cycle, advanced analytics, culture embedded, governance adapts proactively to emerging technology and regulation	Annual governance review shows improvement; innovation pipeline active; examination results positive; governance adjusts ahead of regulatory changes

USING THE MATURITY ROADMAP

The roadmap is a planning tool. Look at where you actually are, name the milestones to the next stage, and write a plan your staffing can survive.

Progress will not be linear. Some areas will mature faster than others, and external events—a regulatory examination, a vendor AI incident, a new technology deployment—may accelerate or redirect the plan. The governance committee should review progress against the roadmap at least annually and adjust the plan as circumstances warrant.

The self-assessment indicators in the rightmost column provide objective markers for each stage. When the committee can honestly affirm that all indicators for a stage are met, the bank has achieved that maturity level and can focus on advancing to the next.

A Realistic Perspective on Maturity

Most community banks starting their AI governance journey today will reach the Foundational stage within six months and the Evolving stage within a year. Reaching Established maturity typically takes eighteen months to two years, and Advanced maturity requires a sustained, multi-year commitment. There is no shame in the Foundational stage. The distinction that matters is between a bank that has started and a bank that has not.

17.5 Final Reflections: Governance as Stewardship

This book began with a community bank president who received three pieces of AI-related news in a single morning and was not yet equipped to respond. Seventeen chapters later, that president—and the IT leader, the governance committee, and the board—have the program to respond with confidence.

But the program is only a means to an end, and the end is stewardship.

Community banks are stewards. They are stewards of their customers' deposits, of their borrowers' trust, of their communities' economic vitality, and of a banking model that values relationships over transactions. Artificial intelligence is a powerful tool, and like all powerful tools, it can serve stewardship or undermine it. The difference is governance.

A community bank that governs AI well protects its customers from unfair or opaque decisions. It ensures that the efficiency gains from AI are reinvested in the community—in more accessible services, in stronger relationships, in more responsive lending. It maintains the human judgment and personal accountability that customers trust. And it demonstrates to regulators, to the community, and to its own employees that the bank takes its responsibilities seriously.

A community bank that neglects AI governance risks the opposite: decisions that customers cannot understand, biases that erode fairness, data practices that betray trust, and a creeping automation that replaces the human connection at the heart of community banking.

The choice between these futures will be determined by the governance decisions that bank leaders make today.

The principles are clear: proportionality, transparency, fairness, accountability, privacy, and safety. The regulatory expectations are established. The program is are in your hands. The maturity roadmap shows the path.

What remains is to begin—or to continue—the work. Community banks have always been defined by doing right by the people they serve. AI governance is the newest expression of that old promise.

Begin with the inventory you can build this month, the policy you can adopt this quarter, and the committee you can convene next week. The program will grow from there—programs always do, once someone owns them.

Chapter Summary

This chapter has positioned your governance program to evolve with the technology it governs:

- Emerging technologies—agentic AI, multimodal AI, embedded AI, synthetic data—will expand the governance challenge, but principles-based governance provides the flexibility to adapt.
- The regulatory frontier is moving toward greater specificity and rigor; proactive governance today reduces the cost of future compliance.
- Industry collaboration—associations, peer networks, shared services—accelerates maturity and reduces costs for individual banks.
- The maturity roadmap—four stages beyond the Initial starting point: Foundational, Evolving, Established, Advanced—provides a realistic multi-year path.
- AI governance is an act of stewardship—an expression of the values that have always defined community banking, applied to a powerful new technology.

Map Where You Are. Plan Where You're Going

The maturity model in this chapter becomes a planning tool with the right format. The AI Governance Toolkit includes the stage-gate AI Governance Maturity Assessment—an Excel scorecard inside the Board AI Reporting Pack—to set your baseline, define your next-stage targets, and present the result to the board as your annual AI governance plan. See the Toolkit page at the back of this book.

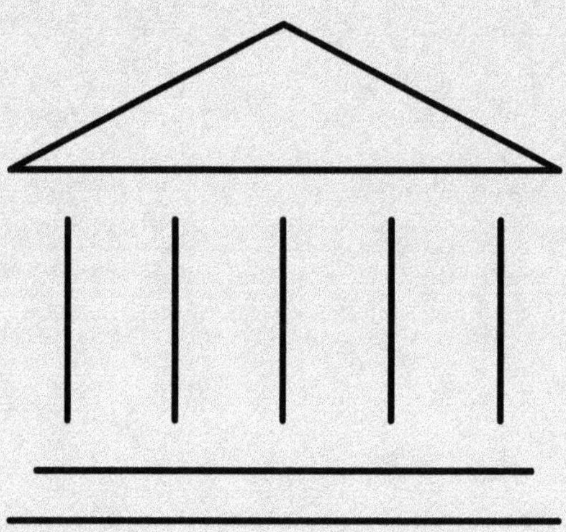

WHAT FOLLOWS

The program is now complete; the rest of the book shows it in use. The eight case studies that follow pressure test the program—fictionalized situations drawn from real community banking experience, each with its lesson drawn out. The appendices then show what good looks like filled in: every core document in this book, completed for a fictional community bank and ready to be adapted as your own, with the regulatory citations behind every chapter collected at the end.

"Example is the school of mankind, and they will learn at no other."

— EDMUND BURKE —

Case studies: Governance in Action

FICTIONALIZED SCENARIOS DRAWN FROM REAL COMMUNITY BANKING SITUATIONS

The following case studies illustrate AI governance principles in action. Each scenario is fictionalized but drawn from situations that community banks actually encounter. They can be used in governance committee discussions, adapted for board education sessions, or incorporated into employee training exercises.

Note: All bank names, individual names, and specific details are fictional. Any resemblance to actual institutions is coincidental.

CASE STUDY 1:

The Agent That Kept Going

Relevant to: Chapter 7 (Section 7.6), Chapter 11 (Sections 11.1–11.3), Chapter 12 (Section 12.5)

Case Study

Maple Creek Community Bank—$720 million in assets, two-state footprint, 110 employees

The Situation

Maple Creek's deposit operations team piloted an agentic assistant inside its document-processing platform—a capability the vendor had added in a quarterly release. The agent handled returned-mail exceptions end to end: it read the returned-mail flag, located the most recent customer-uploaded document containing an address, updated the customer record, reissued held statements, sent a confirmation email, and logged the case closed. In its first month it cleared a backlog of 300 exceptions, and the team stopped checking its work.

On a Wednesday at 9:14 a.m., the agent processed an exception for a customer whose record was indexed to a utility bill the customer's loan co-signer had uploaded during a lending application. The agent extracted the co-signer's address, updated the customer's record, and queued two months of held statements for reissue to an address belonging to someone else. At 9:15 it emailed a change-of-address confirmation. By 9:40 it had processed four similar exceptions using the same flawed document logic. The first complaint surfaced Friday afternoon—a customer asking why the bank had "moved" him. By then, statements for three customers were already in the mail stream to the wrong addresses.

The Response

The operations manager used the agent's kill switch Friday at 4:55 p.m.—the first time it had been used outside a demo—and the queue reverted to manual processing. Over the weekend, the team reconstructed every action from the agent's logs: 412 exceptions processed in five weeks, five wrong-address updates, three resulting in misdirected statements. Monday morning the incident process took over: counsel assessed notification obligations for the three disclosures, the operations manager called each affected customer herself, the bank made them whole, and the governance committee received the log reconstruction at an emergency session under its escalation protocol.

The post-incident review located the root cause in how the agent was deployed rather than in the software itself: it had gone live through a vendor release with no risk assessment, no autonomy cap, and no approval gate on address changes—a bright-line customer-data action under the bank's own draft policy. The agent returned to service six weeks later with a human gate on every address change, a daily action-log review, and a kill-switch test on the calendar each quarter.

The Lesson

An agent's first failure is rarely dramatic; it is quiet, plausible, and repeated. Maple Creek's losses stayed small because the chain was short and the logs were complete. The controls that contained the incident—kill switch, full action logs, escalation protocol—made the difference between an incident the bank could fully reconstruct by Monday and a pattern it would have discovered months later, one complaint at a time. The controls it lacked—pre-deployment assessment, autonomy caps, gates on bright-line actions—are the ones Chapter 11 asks you to build before the pilot, because the time to install them is before the agent clears its first backlog.

CASE STUDY 2:

The Vendor Upgrade That Nobody Noticed

Relevant to: Chapter 4 (Section 4.4), Chapter 8 (Section 8.1), Chapter 9 (Sections 9.2, 9.4)

Case Study

Prairie First Bank—$1.2 billion in assets, three-state footprint, 140 employees

The Situation

During a routine quarterly update, Prairie First's core processing vendor activated a new AI-powered fraud detection module. The release notes mentioned the feature in a single paragraph buried on page fourteen of a twenty-page document. No one at the bank read the full release notes. The AI module began scoring every debit card transaction in real time, declining transactions it flagged as potentially fraudulent.

Within the first week, the bank received seventeen customer complaints about legitimate transactions being declined. A local business owner—a longtime customer and personal friend of a board member—had a $4,200 equipment purchase declined at a farm supply store. The branch manager could not explain why. The IT department did not know the AI module existed.

The Response

The IT manager contacted the vendor, confirmed the AI module had been activated by default, and requested immediate documentation. The vendor provided a two-page summary that answered none of the bank's governance questions. The bank temporarily reverted to the previous fraud detection rules while conducting a proper risk assessment.

The incident bought three governance improvements: a formal review of vendor release notes for AI components before any update goes live, AI-specific contractual provisions in the vendor management program, and a standing vendor-AI agenda item at the governance committee.

The Lesson

Vendor-embedded AI is the most common and least visible AI risk in community banking. If your governance program does not include a process for identifying AI features in vendor updates, you may already be using AI in regulated activities without knowing it. The contractual provisions in Chapter 9 and Appendix C are written for this scenario.

CASE STUDY 3:

When the Model Learned the Wrong Lesson

Relevant to: Chapter 4 (Section 4.2), Chapter 5 (Section 5.2), Chapter 9 (Section 9.2), Chapter 15 (Section 15.1)

Case Study

River Valley National Bank—$850 million in assets, serving a mid-size metropolitan area, 120 employees

The Situation

River Valley had been using a vendor-supplied AI credit scoring model for eighteen months when a new compliance officer, ninety days into the job, ran the bank's first disparate impact analysis. The results were alarming: the AI model's denial rate for applicants in predominantly minority census tracts was 34% higher than for comparable applicants in majority-white tracts, even after controlling for income, credit history, and debt-to-income ratio.

The vendor's model used geographic and behavioral data features that, while not explicitly including race or ethnicity, served as proxies for protected characteristics. The vendor had not disclosed this proxy risk in its documentation, and the bank had not asked.

The Response

The compliance officer escalated the finding to the CEO and the newly formed AI governance committee. The bank engaged a third-party fair lending consultant to validate the analysis and quantify the impact. The findings confirmed statistically significant disparate impact that could not be fully explained by legitimate credit risk factors.

The bank took several immediate actions: notified legal counsel, initiated a review of all applications affected by the model during its deployment period, contacted the vendor to demand bias testing results and model documentation, and began evaluating alternative models. The governance committee implemented ongoing quarterly fair lending monitoring for all AI lending systems and added bias testing to the vendor due diligence questionnaire.

The bank self-reported the finding to its regulator, demonstrating proactive governance and good faith. The examiner's response, while serious, recognized the bank's self-identification, remediation efforts, and newly implemented controls.

The Lesson

AI bias is measurable and discoverable—but only if you test for it. River Valley's story would have ended very differently if the disparate impact had been discovered by an examiner or a customer lawsuit rather than by the bank's own compliance function. The fair lending testing methodology in Chapter 5 and the vendor questions in Chapter 9 exist so this story ends at a compliance officer's desk instead of an examiner's report. And although the CFPB's 2026 rule removing disparate impact from Regulation B (effective July 21, 2026) shifted the federal posture, an outcome like this one remains actionable under the Fair Housing Act, under state fair-lending laws, and as reputational and safety-and-soundness risk—which is why the case for testing is as strong as ever.

CASE STUDY 4:

The Board That Asked the Right Question

Relevant to: Chapter 12 (Section 12.3), Chapter 13 (Section 13.1, 13.6)

Case Study

Summit Community Bank—$500 million in assets, rural mountain community, 65 employees

The Situation

Summit's board had been receiving quarterly AI governance reports for two quarters. The reports were well formatted, the dashboard was green across all systems, and the committee appeared to be functioning effectively. During the third quarterly presentation, a board member who had recently attended a community banking conference asked a simple question: "What happens if our lending AI fails on a Saturday morning when nobody is in the office?"

The room went quiet. The IT manager acknowledged that while the bank had business continuity plans for system outages, there was no specific fallback procedure for AI system failures. If the AI credit scoring model produced erroneous results over a weekend, automated pre-approvals through the digital banking platform could issue incorrect decisions to customers for up to 48 hours before anyone noticed.

The Response

The question prompted a comprehensive review of business continuity provisions for all High- and Critical-tier AI systems. The IT manager developed specific fallback procedures for each system, including automatic alerts triggered by anomalous AI output patterns, documented manual override procedures accessible to weekend duty officers, defined thresholds at which AI systems would automatically revert to rules-based fallbacks, and an updated incident response plan with AI-specific scenarios.

The governance committee added business continuity assessment as a required element of every High- and Critical-tier risk assessment, and the board requested a tabletop exercise simulating an AI failure scenario—which revealed two additional gaps in the response plan.

The Lesson

Effective board governance runs on the willingness to ask practical, scenario-based questions that test whether governance works in the real world—no technical credential required. The board member's question—drawn from the reference card questions in Appendix H—uncovered a material gap that the governance committee and IT team had overlooked. This is exactly what board oversight is for.

CASE STUDY 5:

The Chatbot That Gave Financial Advice

Relevant to: Chapter 5 (Section 5.4), Chapter 7 (Section 7.6), Chapter 15 (Section 15.3)

Case Study

Lakeshore Savings Bank—$750 million in assets, suburban market, 95 employees

The Situation

Lakeshore deployed a vendor-supplied AI chatbot on its digital banking platform to handle routine customer inquiries—branch hours, account balances, transaction history, and basic product information. The chatbot worked well for three months. Then a customer asked the chatbot: "Should I take money out of my savings to pay off my credit card?"

The chatbot responded with a detailed analysis recommending that the customer liquidate their emergency savings to pay down credit card debt, citing the interest rate differential as justification. The customer followed the advice. Three months later, a medical bill arrived, and the emergency fund that should have absorbed it was gone. The customer complained to the bank, and the complaint escalated to the branch manager, then to the compliance officer.

Investigation revealed the pattern: the chatbot had been dispensing personalized financial recommendations for weeks—advice that could be construed as financial planning, which the bank was not licensed to provide and which the chatbot was never intended to deliver.

The Response

The bank immediately restricted the chatbot's response scope, implementing hard guardrails that prevented it from providing any personalized financial recommendations. The vendor worked with the bank to add explicit disclaimers and automatic escalation to a human banker for any question that could involve financial advice.

The governance committee conducted a post-incident review that produced three changes: all customer-facing AI systems now undergo monthly output sampling to detect scope creep, the chatbot's initial disclosure was strengthened to clearly identify it as an AI providing general information only, and the bank's AI incident response policy was updated to include "scope creep" as a defined incident category.

The Lesson

AI chatbots do not understand the boundaries of their intended role. They respond to whatever they are asked, and their responses can drift into territory that creates compliance, legal, and customer harm risks. If you deploy a customer-facing AI, you must define both what it should do and what it must not do—and you must monitor continuously to ensure it stays within those boundaries. Chapter 15's guidance on chatbot governance exists because scenarios like Lakeshore's happen more often than the vendors acknowledge.

CASE STUDY 6:

Two Banks, One Examiner

Relevant to: Chapter 4 (Section 4.7), Chapter 17 (Section 17.4)

Case Study

Two community banks—similar size ($600-700 million), same examiner, same examination cycle

The Situation

An FDIC examiner was assigned to examine two community banks of similar size in the same quarter. Both banks used approximately the same number of AI systems—a vendor-supplied credit scoring model, an AI fraud detection module in the core processor, a digital banking chatbot, and various employees using generative AI tools.

At Bank A, when the examiner asked about AI governance, the CEO responded: "We rely on our vendors to manage that." There was no AI inventory, no policies addressing AI use, no documented risk assessments, and no evidence of board or committee oversight. The IT manager was aware of some AI systems but had not documented them. Employees confirmed using ChatGPT for work with no guidance or restrictions.

At Bank B, the examiner asked the same question. The IT manager reached for a binder: a documented AI inventory (using Appendix F), risk classifications for every system, board-approved AI governance and generative AI policies (based on Appendices A and G), quarterly committee meeting minutes showing substantive discussion, risk assessments for all High- and Critical-tier systems (using Appendix B), and the most recent quarterly board report (using Appendix D). The governance program had been in place for only eight months.

The Response

Bank A received multiple examination findings related to model risk management, third-party risk oversight, and information security governance, including a Matter Requiring Board Attention (MRBA) tied to its highest-stakes systems: an ungoverned credit-scoring model and an unmonitored fraud module made the missing governance a material-risk matter, not a paperwork exercise. The bank faced an expensive, urgent remediation effort under regulatory pressure.

Bank B received no AI-related findings that cycle. The examiner looked past the binder—sampling the credit model's monitoring reports and the chatbot's complaint log to confirm the program was lived-in—and noted the governance program as a positive indicator of proactive risk management. In the exit interview, the examiner commented that the bank's governance program, while still maturing, demonstrated the awareness, documentation, and oversight structure the agency expects.

The Lesson

Technology, budget, and staff size were nearly identical at the two banks; governance made the difference. Bank B's program was eight months old, built with limited resources, and far from perfect. But it had the essential elements: inventory, classification, policies, committee oversight, risk assessments, and board reporting. That was enough—because the program was lived-in, and the examiner checked. The documents Bank B used are the ones in this book's appendices; the discipline that made them credible was the bank's own. The choice between Bank A's outcome and Bank B's outcome is yours to make.

CASE STUDY 7:

The Employee Who Fed the Model

Relevant to: Chapter 7 (Section 7.7), Chapter 10 (Section 10.7), Chapter 16 (Section 16.1)

Case Study

Ridgeview Community Bank—$450 million in assets, suburban market, 70 employees

The Situation

A commercial loan officer at Ridgeview Community Bank was preparing a complex credit package under a tight deadline. To speed up the process, he pasted three years of the borrower's tax returns, personal financial statements, and a draft credit memo into a free-tier generative AI chatbot, asking it to summarize the financial position and draft a recommendation.

The output was impressive—a well-organized summary that would have taken hours to write manually. The loan officer shared the approach with two colleagues, who began doing the same with their loan files. Within a month, at least fifteen borrowers' confidential financial data had been uploaded to a third-party AI service with no data processing agreement, no encryption guarantees, and terms of service that permitted the provider to use uploaded content for model training.

How it was Discovered

During a routine network review, the IT manager noticed unusual outbound data volume to an unfamiliar domain. A brief investigation revealed the pattern: multiple lending staff were copying sensitive customer data into a public AI tool. The IT manager escalated to the compliance officer, who recognized the potential GLBA and state privacy law implications immediately.

The Response

The bank assembled an incident response team within 24 hours. The immediate actions included blocking the AI service at the network level, cataloging exactly which customer records had been exposed, consulting outside counsel on notification obligations, and briefing the board. The bank's legal counsel determined that because the uploaded data included Social Security numbers and financial account details, the exposure triggered notification requirements under the bank's state data breach notification law.

Within two weeks, the bank adopted a Generative AI Acceptable Use Policy (modeled on Appendix G of this book), deployed a data loss prevention tool to flag outbound transfers of sensitive patterns like SSNs, conducted mandatory all-staff training on approved and prohibited AI uses, and established a pre-approved list of enterprise AI tools with appropriate data processing agreements. The loan officer was not terminated—he was retrained, because he had been trying to do good work faster with no policy to guide him. But the incident became a centerpiece of the bank's training program.

The Lesson

The whole breach took a helpful employee and an open chatbot window. The loan officer was productive, resourceful, and completely unsupervised in his use of a powerful technology. That combination—diligence without guardrails—is exactly what governance is designed to address. A Generative AI Acceptable Use Policy (Chapter 7, Section 7.7) would have prevented this entirely—by channeling AI use through approved tools with proper data protections. If your bank does not have this policy in place today, Appendix G shows you exactly what that policy looks like, and the AI Governance Toolkit provides it as an editable file you can complete and adopt this week. Ridgeview found the pattern through a network review; the next bank may hear about it first from a teller, a customer, or an examiner.

CASE STUDY 8:

The Voice That Wasn't

Relevant to: Chapter 15 (Sections 15.8, 15.9), Chapter 16 (Section 16.1)

Case Study

Cedar Ridge Bank—$780 million in assets, small-metro market, 100 employees

The Situation

On a Thursday afternoon, a treasury-services associate at Cedar Ridge received a call from a number that displayed as the bank's largest commercial customer—a regional construction company. The voice on the line was unmistakably the company's CFO: the same cadence, the same slight regional accent, even a reference to the golf outing the two had attended the month before. The "CFO" explained that a closing had moved up, that the company's controller was traveling, and that a $480,000 wire to a new supplier needed to go out before end of day. He followed up with an email—from what appeared to be the company's domain—attaching wiring instructions on the company's letterhead.

Everything looked and sounded right. It was not. The voice had been cloned from the CFO's quarterly-earnings webcast; the email domain was a look-alike with a single transposed letter; the letterhead had been generated from public marketing materials.

What Saved the Bank

Cedar Ridge had adopted a callback-verification policy after reading about deepfake fraud in an industry briefing. The policy was simple and absolute: any wire over $100,000 involving new or changed beneficiary instructions required a callback to a phone number already on file for the customer—never a number supplied in the request. The associate, despite being entirely convinced by the voice, followed the procedure. She called the CFO's known direct line. The real CFO answered, confused: there was no closing, no new supplier, no wire. The fraud collapsed in a ninety-second phone call.

The Response

The bank filed a Suspicious Activity Report using FinCEN's deepfake-fraud keyword, alerted the customer, and shared a sanitized account of the attempt with its peer network. The governance committee used the near-miss to harden controls further: callback verification was extended to all beneficiary changes regardless of dollar amount, the bank added a brief deepfake-awareness module to its annual fraud training, and treasury-services staff were given explicit authority—and explicit cover—to slow down any payment request, no matter how senior or convincing the person on the other end appeared to be.

The Lesson

AI has severed the link between sounding real and being real. A familiar voice, a known face on a video call, a letterhead, even a caller-ID match—none of these are proof of identity anymore, and a bank's controls must be designed on that assumption. The defense that worked cost nothing and predated the threat: a simple, mandatory, out-of-band verification step, and a culture that empowered a junior employee to use it against a senior-sounding request. Cedar Ridge's story is a success only because someone followed a boring procedure when every human instinct said the call was genuine. That is what good governance feels like in the moment: unglamorous, slightly awkward, and exactly right.

APPENDICES

Policies, Tools & Resources

The following appendices contain completed worked examples of the program's core documents. Each appendix has been filled in using a fictional bank—Heartland Community Bank, a $650 million rural Midwest institution with 85 employees—to show the standard filled in. The AI Governance Toolkit provides the program's documents as blank, editable files your team can complete and deploy for your own institution. These pages are, in effect, the binder Bank B reached for in Case Study 6.

Appendix A: AI Governance Policy

A comprehensive enterprise AI governance policy, shown as adopted by a fictional community bank. (Editable version in the Toolkit)

This is the worked example of the AI Governance Policy from Chapter 7 (Section 7.1)—the umbrella document of the seven-policy suite. Its twelve sections incorporate the governance structure, risk tiers, and acceptable-use provisions described in that chapter; the remaining suite policies operate as companion documents.

---

1. Purpose

This policy establishes the governance program for the use of artificial intelligence at Heartland Community Bank. It defines the principles, roles, responsibilities, and processes for the responsible development, acquisition, deployment, monitoring, and retirement of AI systems across the organization.

2. Scope

This policy applies to all AI systems used by Heartland Community Bank, including systems developed internally, acquired from vendors, embedded in third-party platforms, and accessed by employees through external tools. It applies to all employees, officers, directors, contractors, and third parties acting on behalf of the bank.

3. Definitions

Artificial Intelligence (AI): Any system that uses computational techniques to perform tasks that traditionally required human judgment, pattern recognition, or decision-making, including machine learning, robotic process automation, natural language processing, and generative AI.
AI System Owner: The designated senior manager responsible for the governance of a specific AI system throughout its life cycle.
AI Governance Committee: The cross-functional body responsible for overseeing AI governance across the organization, as defined in Section 5.
Risk Tier: The classification assigned to an AI system based on its risk assessment: Minimal, Moderate, High, or Critical.

4. Governing Principles

All AI use at Heartland Community Bank shall be governed by the following principles: Proportionality—governance effort shall be commensurate with risk. Transparency—AI decisions shall be explainable to appropriate audiences. Fairness—AI systems shall be tested and monitored for bias. Accountability—every AI system shall have a named human owner. Privacy—customer data shall be protected in all AI processing. Safety—AI systems shall be designed for graceful failure with human override capability.

5. Governance Structure

The AI Governance Committee shall be chaired by a senior officer designated by the CEO (currently the IT Officer) and shall include representatives from IT, Risk/Compliance, and relevant business lines. The Committee shall meet at least quarterly and shall have authority to approve or reject AI use cases, establish governance standards, and escalate matters to the Board. The Board of Directors shall approve this policy, receive quarterly AI governance reports, and set the bank's AI risk appetite.

6. AI Inventory and Risk Classification

All AI systems shall be recorded in the AI inventory maintained by the IT Officer, with oversight by the AI Governance Committee. Each system shall be classified as Minimal, Moderate, High, or Critical risk using the bank's AI Risk Assessment methodology. Governance requirements shall be proportionate to the assigned risk tier.

7. Acceptable Use

Employees shall use only approved AI tools for bank business. Use of unapproved AI tools, including consumer generative AI platforms, for processing bank or customer data is prohibited. Requests to use new AI tools shall be submitted through the AI use case intake process for governance review.

8. Vendor AI Governance

AI systems acquired through vendor relationships shall be subject to AI-specific due diligence, contractual provisions, and ongoing monitoring as defined in the bank's AI Vendor and Third-Party Policy.

9. Monitoring and Validation

All AI systems rated Moderate or above shall be subject to ongoing monitoring with defined KPIs. High and Critical systems shall undergo independent validation on a risk-triggered basis, with validation currency reviewed at least annually for Critical systems, consistent with the bank's monitoring requirements. Monitoring results shall be reported to the AI Governance Committee.

10. Incident Response

AI incidents shall be managed in accordance with the bank's AI Incident Response Policy. Significant incidents shall be escalated to the AI Governance Committee, senior management, and the Board as appropriate.

11. Training

All employees shall receive AI governance training appropriate to their role. Training shall be completed upon hire and refreshed at least annually.

12. Policy Review

This policy shall be reviewed at least annually by the AI Governance Committee and approved by the Board of Directors.

Board Approval Date: January 14, 2026
Next Review Date: January 14, 2027
Policy Owner: Karen Mitchell, Chief Risk Officer

Appendix B: AI Risk Assessment

A completed AI risk assessment evaluating LoanScore Pro, a Critical-tier AI credit decisioning system. (Editable version in the Toolkit)

System Identification

AI System Name: LoanScore Pro—AI Credit Decisioning Module
System Owner: Sarah Chen, Chief Lending Officer
Vendor (if applicable): CoreBanc Technologies (embedded in LOS platform)
Business Function: Consumer and small business lending—credit decisioning
Assessment Date: March 15, 2026
Assessor(s): Mike Torres (IT Officer), Sarah Chen (CLO), Lisa Park (Compliance Officer)

System Description

Purpose / Business Problem Addressed: Generates credit risk scores and recommended approval/denial decisions for consumer and small business loan applications. Supplements traditional credit bureau scoring with broader pattern recognition. Bank has used the prior generation of this model since 2022.
AI Technique(s) Used: Gradient-boosted decision tree ensemble model trained on a national dataset of 2.3 million loan outcomes. Updated quarterly by the vendor.
Data Inputs: Credit bureau data (Equifax, TransUnion), application data (income, employment, DTI), bank account transaction history (12 months), property/collateral data for secured loans. Does NOT use protected class data directly.
Outputs / Decisions Produced: Numeric risk score (300–850), recommended decision (Approve/Refer/Deny), up to 4 adverse action reason codes, confidence indicator (High/Medium/Low).

Users / Affected Parties: Loan officers (direct users), all consumer and small business loan applicants (affected parties). Approximately 2,400 applications scored annually.

Risk Dimension Scoring (1=Minimal, 2=Moderate, 3=High, 4=Critical—higher scores mean higher risk)

RISK DIMENSION	SCORE (1–4)	RATIONALE
Customer Impact	4	Automated credit decisions with direct, material impact on applicants.
Decision Autonomy	3	Renders approve/refer/deny decisions within approved parameters; a loan officer reviews each before action.
Regulatory Sensitivity	4	Core fair-lending (ECOA/FHA) and adverse-action exposure.
Data Sensitivity	3	Uses customer PII and credit-bureau data.
Model Complexity	3	Vendor ML model; interpretable via reason codes, limited feature documentation.

Overall Risk Tier (highest score determines tier): CRITICAL (driven by Customer Impact and Regulatory Sensitivity scores of 4)

Risk Analysis

Key Risks Identified: (1) Disparate impact risk: model uses geographic and behavioral features that may correlate with race/ethnicity. (2) Explainability gap: vendor provides reason codes but limited documentation of feature importance. (3) Concentration risk: 100% of consumer credit decisions rely on this single model. (4) Model drift: vendor updates may shift behavior without notice.

Existing Controls: Loan officer review of all decisions. Four-eye approval for denials over $50K. Quarterly lending report reviewed by compliance. Annual HMDA analysis. Vendor contractual obligation for adverse action reason codes meeting Reg B requirements.

Residual Risk Assessment: MODERATE–HIGH. Existing controls address some risks but lack AI-specific testing. No disparate impact analysis has been conducted on this specific model. Explainability gap could create regulatory and reputational exposure.

Bias / Fair Lending Considerations: Vendor has not provided disparate impact testing results despite request. Geographic features (ZIP code, census tract data) are known proxies for race/ethnicity. Bank has not conducted its own disparate impact analysis on this model. This represents a material gap requiring remediation.

Governance Plan

Monitoring Plan (KPIs, frequency): KPIs: approval/denial rates by product and demographic segment (monthly; non-HMDA segments estimated by proxy methodology under the fair-lending consultant's direction), model score distribution stability (monthly), override rate tracking (monthly), customer complaint analysis (quarterly), HMDA outcomes review (annually). Dashboards reviewed by lending committee monthly and governance committee quarterly.

Validation Plan (method, frequency): Independent disparate impact analysis underway (Midwest Lending Analytics Group, engaged January 2026); complete by Q2 2026. Annual independent model validation thereafter. Vendor to provide updated model documentation annually as part of contract renewal.

Interim Compensating Controls (pending the Q2 2026 analysis): enhanced second review of all denials in majority-minority census tracts; monthly cross-geography denial-rate comparison reported to the Compliance Officer; board-documented acceptance of continued operation during the remediation window; and a look-back review of affected applications, scoped in advance, if the analysis confirms disparity. Suspension of the model in favor of manual underwriting (the documented fallback, at 60% capacity) was evaluated and not elected; that decision and its rationale are recorded for board ratification.

Human Override Mechanism: All model recommendations reviewed by assigned loan officer before decision. Override authority documented. Override reasons logged in LOS. Quarterly override analysis presented to lending committee.

Business Continuity / Fallback Process: If model unavailable: revert to manual underwriting using bank's traditional credit policy and bureau scores. Estimated processing capacity: 60% of normal volume. IT to be notified within 1 hour of model unavailability.

Vendor Governance Requirements (if applicable): Contractual: annual model documentation update, quarterly performance reports, bias testing results (to be negotiated in upcoming contract renewal), 30-day advance notice of material model changes, audit rights.

Approval

Recommendation (Approve / Conditional / Defer / Reject): CONDITIONAL APPROVAL—System may continue in use subject to the following conditions being met within stated time frames.

Conditions (if applicable): (1) Complete independent fair lending analysis by June 30, 2026. (2) Negotiate vendor contract amendment requiring annual bias testing results by contract renewal date (September 2026). (3) Document feature inventory and obtain vendor confirmation that no protected class proxies are intentionally used. (4) Reassess in 90 days post-conditions completion.

Committee Approval Date: March 22, 2026

Board Notification Date (if required): April 2026 quarterly board report

---

Appendix C:
AI Vendor Due Diligence Questionnaire

A completed vendor due diligence assessment for the AI credit decisioning vendor described in Appendix B. The 25 questions are the questions to ask of any AI vendor (from Chapter 9); the header and Overall Assessment block show what a completed evaluation looks like. (Editable version in the Toolkit)

Vendor Name: CoreBanc Technologies
Product / Service: LoanScore Pro—AI Credit Decisioning Module (embedded in LOS platform)
Assessment Date: January 22, 2026
Assessor: Mike Torres, IT Officer / AI Governance Committee Chair

---

Model Transparency (Questions 1–5)

1. What AI or machine learning techniques does your product use, and for what specific functions?
2. What data inputs does the model use, and how were they selected?
3. Can you provide documentation explaining the model's decision logic at a level sufficient for our regulatory obligations?
4. How explainable are the model's individual decisions? Can we generate reason codes or explanatory factors for each output?
5. What are the known limitations of the model, and in what scenarios might it perform poorly?

Bias and Fairness (Questions 6–10)

6. Has the model been tested for disparate impact on protected classes? What methodology was used and what were the results?
7. How frequently is bias testing conducted, and will you share results on an ongoing basis?
8. If bias is detected, what is your remediation process?
9. Does the model use data elements that could serve as proxies for protected characteristics?
10. How do you ensure that model updates do not introduce new biases?

Data Governance (Questions 11-15)

11. What customer data does your AI system process, and where is it stored?
12. Is our customer data used to train or improve models that serve other clients?
13. What happens to our data if we terminate the vendor relationship?
14. What security controls protect data processed by your AI system?
15. How do you handle data retention and deletion for AI training data?

Operations and Resilience (Questions 16-20)

16. How do you validate your AI models, and how frequently?
17. What monitoring do you perform on model performance, and will you share reports?
18. How do you handle model updates? Will we be notified before material changes take effect?
19. What is your incident response process for AI failures or unexpected outcomes?
20. What business continuity provisions exist if your AI system experiences an outage?

Compliance and Governance (Questions 21-25)

21. Do you have a formal AI governance program? Describe its key elements.
22. Will you support our regulatory examination process by providing documentation and responding to examiner inquiries?
23. Are you willing to grant us audit rights specific to your AI systems?
24. How do you stay current with evolving AI regulations, and how do you communicate regulatory changes?
25. Can you provide references from other community bank clients regarding your AI governance practices?

Overall Assessment

Vendor Governance Maturity Rating (Strong / Adequate / Needs Improvement / Inadequate): Needs Improvement

Key Gaps Identified: Vendor declined to provide complete disparate impact testing results, citing competitive sensitivity. No independent third-party model validation report available at time of assessment. No formal AI-specific incident response process documented. Vendor model documentation last updated August 2024.

Compensating Controls Required: Heartland to conduct independent third-party fair lending analysis by June 30, 2026 (engaged: Midwest Lending Analytics Group). Quarterly performance reports under SLA. Heartland to develop AI-specific incident response procedures internal to the bank. Vendor required to provide 30 days' notice of material model changes.

Recommendation: APPROVED with the compensating controls above. Vendor relationship to be re-evaluated at contract renewal (September 2026). Concentration risk to be addressed in vendor diversification strategy by Q4 2026.

Appendix D: Board AI Reporting

A completed quarterly board AI governance report illustrating tone, depth, and format. Designed to fit three to five pages. (Editable version in the Toolkit)

HEARTLAND COMMUNITY BANK

Quarterly AI Governance Report

Reporting Period: Q1 2026 (January – March 2026)

---

1. Executive Summary

The bank's AI governance program completed its first full quarter of operation. The AI inventory grew from 18 to 22 systems following identification of four additional vendor-embedded AI features. The model risk profile remains balanced, with two systems classified as Critical (LoanScore Pro and FraudShield), two as High, eight as Moderate, and ten as Minimal. One AI incident occurred this quarter and has been fully resolved.

---

2. AI Inventory Summary

RISK TIER	CURRENT COUNT	ADDED THIS QUARTER	RETIRED THIS QUARTER
Critical	2	0	0
High	2	0	0
Moderate	8	2	0
Minimal	10	2	0
Total	22	4	0

3. High and Critical System Status

Status: Green = performing within parameters;
Yellow = concern under investigation; Red = requires immediate attention.

SYSTEM NAME	RISK TIER	STATUS	KEY NOTES
LoanScore Pro—AI Credit Decisioning	Critical	Yellow	Conditionally approved March 2026. Fair lending analysis underway (target Q2 2026). Vendor: CoreBanc Technologies.
FraudShield—Transaction Monitoring	Critical	Green	Operating within parameters. Annual validation completed February 2026. Vendor: CoreBanc Technologies.
BSA/AML Alert Engine	High	Green	Monthly tuning review. Quarterly back-testing against confirmed cases. Vendor: CoreBanc Technologies.
Digital Banking Chatbot	High	Green	Weekly output sampling initiated Q1 2026 following incident 2026-AI-001. Vendor: Q2 Holdings.

---

4. Incident Summary

Number of AI incidents this quarter: 1
Severity breakdown (High / Moderate / Low): 0 High / 1 Moderate / 0 Low
Open incidents and status: All Q1 incidents resolved as of report date. No open AI incidents.
Lessons learned / governance improvements implemented: Following Incident 2026-AI-001, the bank increased chatbot output sampling cadence from monthly to weekly and added rate-quotation accuracy to the chatbot KPI dashboard. Vendor required to provide source-data refresh confirmation with every chatbot platform update.

---

5. Governance Activity Highlights

Completed risk assessment for LoanScore Pro credit decisioning model (conditionally approved). Adopted Generative AI Acceptable Use Policy (board-approved January 2026). Deployed enterprise Microsoft 365 Copilot platform with DLP controls and mandatory employee training. Implemented vendor update review procedure to identify AI components in software updates. Conducted all-employee AI awareness training (92% completion rate; remaining 8% scheduled for April). Initiated vendor contract review for AI-specific provisions (three contracts under review).

---

6. Regulatory Developments

No examination findings related to AI during Q1. The bank's next FDIC examination is scheduled for Q3 2026. The governance committee is preparing examination materials, including the AI inventory, risk assessments for High- and Critical-tier systems, the seven-policy suite, committee charter, meeting minutes, monitoring reports, and incident documentation.

---

7. Items Requiring Board Action or Awareness

Awareness: LoanScore Pro fair lending analysis is underway with a target completion date of June 30, 2026. The committee will report findings and any recommended actions in the Q2 report. Awareness: Three vendor contracts are under review for AI-specific provisions. Recommendations for contract amendments will be presented to the board as needed. No items require board action at this time.

Report Prepared By: Mike Torres, IT Officer / AI Governance Committee Chair
Report Date: April 3, 2026

Appendix E:
AI Governance Committee Charter

A completed AI Governance Committee Charter showing how a community bank's board would authorize and structure the committee. (Editable version in the Toolkit)

---

1. Purpose

The AI Governance Committee ("Committee") is established to provide cross-functional oversight of artificial intelligence use at Heartland Community Bank. The Committee ensures that AI systems are deployed and managed in accordance with the bank's AI governance policies, risk appetite, regulatory requirements, and ethical principles.

2. Authority

The Committee has authority to approve or reject AI use cases based on risk assessment outcomes; establish, amend, and enforce AI governance policies and standards; require any department, business line, or vendor to provide information necessary for AI oversight; escalate matters to senior management and the Board of Directors; and commission independent reviews, audits, or assessments of AI systems.

3. Scope

This charter covers all AI systems used by the Bank, including vendor-embedded AI, employee use of generative AI tools, and any AI system under development or consideration.

4. Membership

The Committee shall include at minimum: a senior officer designated by the CEO as Chair; the IT leader or CISO; the Chief Risk Officer or compliance designee, and representative(s) from business lines with significant AI use. Additional members may be appointed by the Chair. The Committee may invite subject matter experts, vendors, or external advisors to attend meetings as needed.

5. Meeting Cadence

The Committee shall meet at least quarterly. Ad hoc meetings may be convened by the Chair for urgent matters including AI incidents, high-risk deployment decisions, or regulatory developments requiring immediate attention. A quorum of three members is required for decision-making.

6. Responsibilities

- Maintain and review the AI inventory for completeness and accuracy.
- Review and approve risk assessments for High- and Critical-tier AI systems.
- Oversee AI policy development, implementation, and compliance.
- Monitor AI system performance and risk indicators through the AI risk dashboard.
- Oversee vendor AI due diligence and ongoing monitoring.
- Review AI incidents and ensure appropriate remediation.
- Prepare quarterly AI governance reports for the Board.
- Conduct an annual AI governance effectiveness review.
- Ensure adequate AI governance training across the organization.

7. Reporting

The Committee shall report to the Board of Directors at least quarterly via the quarterly AI Governance Report (the format illustrated in Appendix D of this book). The Committee Chair shall present the report and address board questions. Significant matters shall be escalated to the Board between regular reporting cycles as warranted.

8. Charter Review

This charter shall be reviewed annually by the Committee and approved by the Board of Directors.

Board Approval Date: November 12, 2025
Next Review Date: November 12, 2026

Appendix F: AI Inventory and Classification Worksheet

A field-by-field definition of the inventory record, illustrated with one completed example–Heartland's FraudShield transaction monitoring system –followed by a sample of the full inventory register.
(Editable version in the Toolkit)

FIELD	DESCRIPTION	EXAMPLE
System ID	Unique identifier	AI-2026-001
System Name	Descriptive name	FraudShield–Transaction Monitoring
Vendor / Internal	Source of the AI system	CoreBanc Technologies (Vendor)
Business Function	Department or function served	BSA/AML Compliance
AI Technique	Type of AI (ML, RPA, NLP, GenAI, Agentic)	Machine Learning
System Owner	Named individual with governance responsibility	Compliance Officer
Deployment Date	Date system went live	01/15/2025
Risk Tier	Classification (Minimal/ Moderate/High/Critical)	Critical
Last Risk Assessment	Date of most recent assessment	03/15/2026
Next Assessment Due	Scheduled date for next assessment	03/15/2027
Validation Status	Current/Overdue/Not Required	Current

FIELD	DESCRIPTION	EXAMPLE
Monitoring Cadence	Frequency of monitoring reviews	Monthly
Data Sensitivity	Types of data processed	Customer PII, transactions
Vendor Contract Expiry	Contract end date (if vendor)	12/31/2027
Notes	Additional relevant information	Vendor upgrade scheduled Q3

Recommendation: Maintain this inventory as a spreadsheet with filtering and sorting capability. Review for completeness quarterly. Conduct a comprehensive discovery exercise annually.

Sample Inventory—Heartland Community Bank, March 2026

The sample shows the register's summary columns; the Toolkit's worksheet carries the full 15-field record for each system.

SYSTEM NAME	RISK TIER	STATUS	KEY NOTES
LoanScore Pro—AI Credit Decisioning	Critical	Yellow	Conditionally approved March 2026. Fair lending analysis underway (target Q2 2026). Vendor: CoreBanc Technologies.
FraudShield—Transaction Monitoring	Critical	Green	Operating within parameters. Annual validation completed February 2026. Vendor: CoreBanc Technologies.
BSA/AML Alert Engine	High	Green	Monthly tuning review. Quarterly back-testing against confirmed cases. Vendor: CoreBanc Technologies.
Digital Banking Chatbot	High	Green	Weekly output sampling initiated Q1 2026 following incident 2026-AI-001. Vendor: Q2 Holdings.

SYSTEM NAME	RISK TIER	STATUS	KEY NOTES
Microsoft 365 Copilot (Internal Use)	Moderate	Green	Approved generative AI platform per the bank's Generative AI Acceptable Use Policy (Appendix G). Enterprise license; data residency confirmed.
Treasury Cash-Flow Forecasting Model	Moderate	Green	Internal use only; outputs reviewed by Treasurer before action. Annual validation due Q3 2026.

Appendix G: Generative AI Acceptable Use Policy

A completed Generative AI Acceptable Use Policy as adopted by the fictional Heartland Community Bank—the policy most likely to be tested by daily employee behavior. (Editable version in the Toolkit)

---

1. Purpose

This policy establishes standards for the responsible use of generative artificial intelligence tools by employees of Heartland Community Bank. Generative AI tools include large language models, AI writing assistants, AI coding assistants, image generators, and similar platforms that produce content based on user prompts.

2. Approved Platforms

The following generative AI platforms are approved for use by Heartland Community Bank employees: Microsoft Copilot (Enterprise license, data remains within Heartland Community Bank's Microsoft 365 tenant) and Microsoft 365 Copilot (integrated with the bank's identity management). Approved for: drafting internal communications, summarizing meeting notes, formatting reports, brainstorming marketing concepts, and code-assisted IT scripting in approved sandboxes. Use of any generative AI platform not on this list for bank business is prohibited. Requests to add platforms to the approved list shall be submitted to IT Officer / AI Governance Committee.

3. Prohibited Uses

The following uses of generative AI are strictly prohibited:

- Entering customer personally identifiable information (PII), account numbers, Social Security numbers, or other non-public personal information into any generative AI platform.
- Using generative AI output as the sole basis for credit decisions, regulatory filings, legal opinions, or customer advice.
- Generating communications that impersonate specific individuals or misrepresent AI-generated content as human-authored in contexts where disclosure is required.

- Using generative AI to circumvent any bank policy, security control, or compliance requirement.
- Uploading confidential bank documents, strategic plans, board materials, examination reports, or SARs and SAR-supporting investigation materials to any generative AI platform.

4. Permitted Uses with Conditions

The following uses are permitted subject to the stated conditions:

- Internal drafting: Generative AI may be used to draft internal communications, reports, and analysis, provided all output is reviewed and verified by a human before distribution.
- Research and brainstorming: Generative AI may be used for research, idea generation, and analysis, provided all facts and citations are independently verified.
- Customer communication drafting: Generative AI may assist in drafting customer communications, provided a qualified human reviews the final version for accuracy, completeness, regulatory compliance, and appropriate tone before sending.
- Code and technical assistance: IT staff may use approved AI coding assistants, provided all generated code is reviewed, tested, and validated before deployment.

5. Data Handling

Only the following data categories may be used as inputs to approved generative AI platforms: publicly available information, de-identified aggregate data, the employee's own draft text, and general bank policies or procedures that are not confidential. All other data categories require prior approval from Compliance Officer.

6. Output Verification

All generative AI output used in bank business must be reviewed by a qualified human. Generative AI can produce plausible but incorrect information. Employees are personally responsible for the accuracy and appropriateness of any content they use, regardless of its source.

7. Training

All employees with access to approved generative AI platforms shall complete generative AI training before first use and refresher training annually thereafter.

8. Reporting

Employees shall report to IT Officer / AI Governance Committee any discovery of unapproved generative AI use, any instance where generative AI produces output that appears harmful, biased, or inappropriate, and any suspected data leakage through a generative AI platform. Reports may be made without fear of retaliation.

Policy Effective Date: January 14, 2026
Policy Owner: Karen Mitchell, Chief Risk Officer
Next Review Date: January 14, 2027

Appendix H: Board Member AI Question Reference Card

A one-page reference for board meetings. Print and keep with board materials.

AI GOVERNANCE—QUESTIONS FOR BOARD MEMBERS

The AI Landscape

- How many AI systems are we using, and how does that compare to last year?
- Which systems carry the highest risk classification, and why?
- Are any AI systems operating outside our governance program?

Risk and Performance

- Have any AI systems experienced problems or unexpected results this quarter?
- What are our most significant AI risks, and what is management doing about them?
- When were our high-risk AI systems last tested for bias? What were the results?

Governance Effectiveness

- Is our governance keeping pace with our AI adoption?
- Are governance activities—assessments, validations, monitoring—being completed on schedule?
- What did our last examination reveal about AI governance?

Strategy and Ethics

- How is AI supporting our strategic plan and our community mission?
- Are there AI applications we should pursue or reconsider?
- Are we confident our AI systems treat all customers fairly?

Vendor and Resilience

- How dependent are we on any single AI vendor?
- What happens if our most critical AI system fails tomorrow?
- Are our vendor contracts providing adequate transparency and audit rights?

Appendix I: Regulatory Reference Guide

A quick-reference matrix mapping regulations to governance actions, followed by the AI examination request list—the documents an examiner will ask for, mapped to the chapters that produce them.

REGULATION / GUIDANCE	AI APPLICATION	REQUIRED GOVERNANCE ACTION	REFERENCE CHAPTER
SR 26-2 (formerly SR 11-7) / OCC Bulletin 2026-13	ML models in credit, fraud, pricing, ALM	Model inventory, validation, documentation, board reporting	Ch. 4, 8, 10
ECOA / Regulation B	AI in credit decisions or pricing	Specific adverse action notices, disparate-treatment testing, fair lending analysis (effects-theory exposure persists through case law, FHA, and state law)	Ch. 4, 5, 15
FCRA (Fair Credit Reporting Act)	AI credit decisions using consumer-report data or credit scores	Adverse action notice naming the bureau; credit-score disclosure (score, range, key factors); risk-based pricing notices	Ch. 15
Fair Housing Act	AI in residential mortgage decisions	Disparate-impact testing, redlining analysis, marketing review	Ch. 4, 5
Interagency AVM Quality-Control Rule (2024)	Automated valuation models in covered mortgage decisions	QC policies: estimate confidence, data-manipulation protection, conflict-of-interest avoidance, sample testing, nondiscrimination	Ch. 15
BSA / AML	AI transaction monitoring, SAR processes	Model validation, tuning documentation, human SAR decision, false positive analysis	Ch. 4, 8, 15

REGULATION / GUIDANCE	AI APPLICATION	REQUIRED GOVERNANCE ACTION	REFERENCE CHAPTER
OCC 2023-17 (formerly 2013-29) / FDIC FIL-29-2023	All vendor-supplied AI systems	Due diligence, contractual provisions, ongoing monitoring, contingency planning	Ch. 4, 7, 9
GLBA / Regulation P	AI systems processing customer data	Data governance, access controls, privacy notices, data flow mapping, DLP	Ch. 4, 10
UDAP / UDAAP	Customer-facing AI (chatbots, recommendations)	Output monitoring, disclosure, complaint analysis, escalation rules	Ch. 4, 15
State Privacy Laws	AI processing resident data	Automated decision-making disclosure, data minimization, consumer rights	Ch. 4, 17
Employment Laws	AI in hiring, performance evaluation	Bias testing, adverse impact analysis, documentation	Ch. 3
HMDA / Regulation C	AI use in mortgage underwriting, pricing, and reporting	Accurate data collection and reporting; fair lending integration when AI influences mortgage decisions; documentation of any AI features that affect HMDA-reportable data points.	Ch. 15; App. B

The AI Examination Request List

When AI comes up in an examination, the request list is concrete. The items below are the documents an examiner is most likely to ask for, each mapped to the part of this book that produces it. A bank that can assemble this list in a week is ready for the conversation.

1. AI system inventory with risk classifications
   —Chapter 8; Appendix F.
2. AI governance policy and generative AI acceptable use policy, with board approval dates
   —Chapter 7; Appendices A and G.
3. Governance committee charter, membership, and the past year of minutes
   —Chapter 6; Appendix E.
4. Risk assessments for High- and Critical-tier systems
   —Chapter 8; Appendix B.
5. Model validation reports, or vendor validation documentation, for credit and BSA/AML models
   —Chapters 4 and 10.
6. Fair lending testing results and methodology for AI-influenced lending decisions
   —Chapter 5.
7. Vendor due diligence files and AI contract provisions for significant AI vendors
   —Chapter 9; Appendix C.
8. Board AI reports for the past four quarters
   —Chapter 13; Appendix D.
9. AI incident log and resolution documentation—
   Chapters 7 and 12.
10. Monitoring reports and KPI dashboards for High- and Critical-tier systems, most recent four quarters
    —Chapter 8 (Section 8.4) and Chapter 12.

11. Internal audit reports and remediation tracking for AI-related coverage
—Chapter 14.

12. Training records and policy attestations
—Chapter 16.

The regulatory matrix above shows which regulation drives each request; the chapters show how to produce the artifact. The Toolkit's evidence binder assembles the full set.

Appendix J:
Glossary of AI and Governance Terms

Plain-language definitions for key terms used throughout this book.

TERM	DEFINITION
Adversarial Attack	A deliberately crafted input designed to cause an AI model to produce incorrect outputs.
Adverse Action Notice	A required notification under ECOA and Regulation B informing an applicant when a credit decision is denied, terminated, or made on less favorable terms, identifying the principal reasons. Applies fully to decisions produced or recommended by AI systems.
Adverse Impact Ratio (AIR)	A fair lending screening metric: the favorable-outcome rate of a protected group divided by that of the reference group. An AIR below 0.80—the four-fifths rule, borrowed from federal employment guidance—is the conventional threshold for investigation. It is a screening convention that triggers investigation; a flagged result calls for analysis before any conclusion.
Agentic AI	AI systems capable of autonomously planning and executing multi-step tasks with minimal human intervention.
Algorithm	A set of rules or instructions that a computer follows to solve a problem or make a decision.
Artificial Intelligence (AI)	Technology that enables computers to perform tasks that traditionally required human judgment, pattern recognition, or decision-making.
Backtesting	Applying an AI model to historical data with known outcomes to evaluate how well the model would have performed.
Bias (in AI)	Systematic errors in AI outputs that produce unfair or discriminatory outcomes for certain groups.
Black Box	An AI model whose internal decision logic cannot be readily interpreted or explained.

TERM	DEFINITION
Citizen Developer	A business user who builds automations, applications, or AI agents using no-code platforms, without going through IT. Staff-built automations can bypass intake and inventory, multiplying the shadow AI problem.
Concentration Risk	The risk arising from heavy reliance on a single vendor, technology, or counterparty whose failure or disruption could materially affect bank operations. In AI governance, often associated with dependency on a single AI vendor, model provider, or cloud platform.
Data Lineage	The documented trail of data from its source through any transformations to its use in an AI system.
Data Loss Prevention (DLP)	Technology and processes that detect and prevent unauthorized transmission of sensitive data outside the organization, including to external AI tools.
Data Poisoning	Manipulating the training data used by an AI model to cause it to learn incorrect patterns.
Disparate Impact	When a facially neutral practice disproportionately affects a protected class without legitimate business justification.
Effective Challenge	The constructive questioning of assumptions, methodologies, and conclusions by parties with sufficient knowledge, authority, and independence. A core function of the governance committee in challenging AI risk assessments, monitoring reports, and new use case proposals.
Explainability	The ability to describe how an AI system reaches its decisions in terms understandable to the intended audience.
Fair Lending	The regulatory framework prohibiting discrimination in credit decisions on the basis of protected characteristics. Anchored in ECOA, Regulation B, and the Fair Housing Act. Applies to all credit decisions, including those produced or recommended by AI systems.
False Positive	An alert or flag generated by an AI system for a case that, upon investigation, is not actually problematic.

TERM	DEFINITION
Generative AI	AI systems that create new content—text, images, code, audio—based on user prompts. Includes large language models like ChatGPT.
Hallucination	When a generative AI system produces plausible-sounding but factually incorrect or fabricated output.
Human-in-the-Loop	A system design in which a human reviews, validates, or can override AI decisions before they take effect.
ISO/IEC 42001	The first certifiable international standard for an AI management system, published in December 2023. It sets requirements for establishing, operating, and continually improving AI governance across an organization. For community banks, it is most useful as a signal in vendor due diligence—a vendor certified to ISO/IEC 42001 has had its AI governance independently audited—rather than as a standard the bank itself must adopt.
Kill Switch	A mechanism for immediately halting an AI system's operation when it produces harmful, erroneous, or unauthorized outputs.
KPI (Key Performance Indicator)	A measurable metric used to track AI system performance over time.
Machine Learning	A subset of AI in which systems learn patterns from data to make predictions or decisions without being explicitly programmed.
Materiality	The significance of a risk relative to the institution's size, complexity, and risk profile. SR 26-2 introduces materiality-based oversight, calibrating governance intensity to a system's potential impact rather than applying uniform requirements to all models.
Model Drift	Gradual deterioration in an AI model's performance over time due to changes in data patterns or the business environment.
Model Risk Management	The framework of policies, procedures, and controls applied to statistical and AI models throughout their life cycle—development, validation, deployment, monitoring, and retirement. Governed in U.S. banking by SR 26-2 and OCC Bulletin 2026-13 (formerly SR 11-7 and OCC Bulletin 2011-12).

TERM	DEFINITION
Model Validation	An independent assessment of whether a model is conceptually sound, performs as intended, and remains fit for use—the core discipline of model risk management under SR 26-2.
Multimodal AI	AI systems that process multiple data types simultaneously, such as text, images, and audio.
Natural Language Processing (NLP)	AI techniques that enable computers to understand, interpret, and generate human language.
NIST AI Risk Management Framework (AI RMF)	A voluntary framework from the National Institute of Standards and Technology (NIST AI 100-1, January 2023) for managing AI risk, organized around four functions: Govern, Map, Measure, and Manage. Examiners, auditors, and vendors increasingly treat it as a common reference point. Appendix L maps this book's governance program to the four functions.
Process Mining / Task Mining	Techniques that analyze how work actually happens—process mining reads system event logs; task mining records and classifies on-screen employee activity—to map workflows and identify automation candidates. Discovery recordings can capture customer NPI, raising data governance questions before any automation is built.
Prompt Injection	Crafting inputs to a generative AI system designed to bypass safety controls or extract confidential information.
Proportionality	The principle that governance effort should be commensurate with the level of risk an AI system presents.
Proxy Variable	A data element that correlates with a protected characteristic and may serve as an indirect basis for discrimination.
RACI Matrix	A matrix assigning Responsible, Accountable, Consulted, and Informed roles for governance activities.
Risk Appetite	The level of risk an organization is willing to accept in pursuit of its strategic objectives.
Risk Tier	The governance classification assigned to an AI system based on risk assessment: Minimal, Moderate, High, or Critical.

TERM	DEFINITION
Robotic Process Automation (RPA)	Software that automates repetitive, rules-based tasks such as data entry and report generation.
Scope Creep	The gradual expansion of an AI system's use beyond its originally approved purpose, risk tier, or data inputs without formal reassessment.
Shadow AI	AI tools used by employees outside the organization's knowledge and governance program.
Stress Testing	Evaluating how an AI model performs under adverse, extreme, or unusual conditions.
Synthetic Data	Artificially generated datasets that mimic the statistical properties of real data without containing actual records.
Thin-File Applicant	A loan applicant with limited credit history, for whom traditional scoring models may produce unreliable results and AI alternative data models carry heightened fair lending risk.
Three Lines Model	A governance model in which business functions own risk, risk/compliance provides oversight, and internal audit provides assurance.
Unsupervised Learning	A machine learning approach that finds patterns in data without labeled outcomes—the basis of much anomaly detection, including fraud and AML monitoring. Compare supervised learning, which trains on labeled examples.
Validation (Model)	See Model Validation.
Zero Trust	A security architecture that treats every access request as unverified until proven, regardless of network location. Relevant to AI governance wherever agents and AI services hold credentials of their own.

Appendix K: Data Readiness Scorecard

A completed data readiness assessment for the FraudShield transaction monitoring system, showing how each data quality dimension is scored 1–4 and how compensating controls are documented when gaps exist. *(Editable version in the Toolkit)* The overall score is the lowest individual dimension score, so a single weak dimension sets the overall level. Note the scale direction: in this scorecard a 4 is strong, while in Appendix B's risk scoring a 4 is highest risk. Note also the dates: FraudShield was in production before this assessment—the retrofit condition every first-year program starts in; the Section 10.2 readiness gate applies to new systems going forward.

System Identification

AI System Name:	FraudShield—Transaction Monitoring
Assessment Date:	January 15, 2026
Assessed By:	Mike Torres, IT Officer
Data Sources Evaluated:	Core banking customer master; digital banking platform; FraudShield transaction feed
AI Use Case / Risk Tier:	Real-time transaction fraud monitoring—Critical tier

Data Readiness Scoring

DIMENSION	SCORE (1–4)	EVIDENCE / NOTES	REMEDIATION REQUIRED (IF SCORE 1–2)
Completeness	3	Required fields populated for 98% of transaction records; minor gaps in legacy account metadata	–
Accuracy	3	Transaction data validated daily against core totals; vendor feed reconciliation in place	–
Consistency	2	Customer profile fields differ between core banking and digital banking systems; no single customer master	Manual reconciliation until data integration project (Q3 2026)
Timeliness	4	Real-time transaction feed; latency monitored	–
Accessibility	3	Accessible to monitoring and compliance staff	–
Lineage	3	Source-to-model mapping documented for all critical inputs; transformations logged and reproducible.	–
Governance	3	Data owners assigned; retention schedule applied; classification current	–
Overall Score	2	Lowest dimension (Consistency) governs the overall score	See remediation plan below

Scoring Guide

Score 4 (Strong): Data meets or exceeds requirements for AI deployment. No remediation needed. Score 3 (Adequate): Data is sufficient for deployment with minor documentation gaps. Proceed with normal governance. Score 2 (Gaps Exist): Meaningful data quality issues that could affect AI performance. Deploy only with documented compensating controls and enhanced monitoring. Score 1 (Not Ready): Critical data deficiencies that would compromise AI system reliability. Remediate before deployment.

Assessment Summary

Overall Readiness Score (lowest dimension score): 2—Significant gaps require compensating controls; deployment proceeding under remediation plan

Recommendation:

- ☐ Ready for deployment
- ☑ Deploy with compensating controls
- ☐ Remediate before deployment

Compensating controls required (if applicable): Manual reconciliation of customer profile data across core banking and digital banking systems pending data integration project (Q3 2026); compliance dual-review for any FraudShield alert involving accounts with unreconciled customer data.

Remediation plan and timeline (if applicable): Data integration project (FraudShield → core banking customer master) scheduled Q3 2026; expected to bring Consistency score to 3 (Adequate) by end of Q3 2026 and to 4 (Strong) by Q1 2027 following stabilization.

Sign-Off

Assessed by: Mike Torres, IT Officer Date: 01/15/2026
Reviewed by (Data Owner): Lisa Park, Compliance Officer Date: 01/22/2026
Approved by (AI Governance Committee): January 28, 2026

Appendix L: NIST AI RMF Crosswalk

A mapping of the governance program in this book to the four functions of the NIST AI Risk Management Framework (AI RMF 1.0, January 2023). Examiners, auditors, and AI vendors increasingly treat the NIST framework as a common reference point. This crosswalk lets a community bank show how its program—assembled from the chapters and appendices of this book—already satisfies each NIST function, without adopting a separate framework.

The AI RMF organizes AI risk management into four functions: Govern, Map, Measure, and Manage. Govern is cross-cutting and runs through the other three, which align roughly with the stages of an AI system's life cycle. The mapping below shows where this book delivers each function.

NIST AI RMF FUNCTION	WHAT THE FUNCTION ASKS	WHERE THIS BOOK DELIVERS IT
GOVERN	A culture of AI risk management is cultivated and present.	Six core principles (Ch. 3); the Three Lines Model and the AI Governance Committee (Ch. 6; App. E); the seven-policy suite (Ch. 7; Apps. A, G); board oversight, risk appetite, and reporting (Ch. 13; App. D); role-based training and culture (Ch. 16).
MAP	Context is established and AI risks are identified.	What counts as AI (Ch. 2); the AI inventory (Ch. 8; App. F); the four-tier, five-dimension risk classification (Ch. 8; App. B); vendor and third-party context (Ch. 9; App. C).
MEASURE	Risks are assessed, analyzed, and tracked.	Pre-deployment risk assessment and drift monitoring (Ch. 8); testing and validation—bias, stress, and adversarial (Ch. 10); bias detection and explainability (Ch. 5); the AI risk dashboard (Ch. 12); the Data Readiness Scorecard (App. K).
MANAGE	Risks are prioritized and acted upon.	Risk-tier prioritization and change management (Ch. 8); incident response (Ch. 7, 12); human override, kill switches, and operational fallback (Ch. 10); vendor remediation and concentration-risk planning (Ch. 9); board escalation and reporting (Ch. 13; App. D); the governance maturity roadmap (Ch. 17).

The U.S. Treasury's Financial Services AI Risk Management Framework is organized along NIST AI RMF lines, and the certifiable ISO/IEC 42001 (AI management system) and ISO/IEC 23894 (AI risk management) standards map closely to the same four functions—so a program aligned to this crosswalk is also well positioned for those frameworks as examiner and vendor expectations evolve.

Sources and References

The following sources are organized by chapter. This book is a practitioner guide, and the references reflect the regulatory guidance, legal frameworks, and industry standards that underpin the governance recommendations throughout. Where specific claims are made about regulatory expectations or legal requirements, the source is cited below. Readers are encouraged to consult these primary sources directly, as regulatory guidance evolves and new interpretations emerge over time.

Note: Web URLs are provided where available as of publication. Federal regulatory guidance is available through each agency's public website and through the Federal Financial Institutions Examination Council (FFIEC) at www.ffiec.gov.

Chapter 1: The AI Moment for Community Banks

Board of Governors of the Federal Reserve System, Federal Deposit Insurance Corporation, Office of the Comptroller of the Currency. "Request for Information and Comment on Financial Institutions' Use of Artificial Intelligence, Including Machine Learning." 86 Fed. Reg. 16837 (March 31, 2021).

Independent Community Bankers of America. Policy positions on artificial intelligence and machine learning in community banking. ICBA, accessed 2026.

Conference of State Bank Supervisors. CSBS Artificial Intelligence Advisory Group and related publications on AI in financial services. CSBS, 2024–present.

Chapter 2: The AI You Already Have

This chapter draws on the same sources as Chapter 1; see above.

Chapter 3: Governing AI: The Six Principles

National Institute of Standards and Technology (NIST). "Artificial Intelligence Risk Management Framework (AI RMF 1.0)." NIST AI 100-1. US Department of Commerce, January 2023. See also NIST AI 600-1 (Generative AI Profile, July 2024) for generative-AI-specific risk management guidance, and the draft NIST Cybersecurity Framework AI Profile (preliminary as this book went to press) for mapping AI risks to cybersecurity controls.

Organisation for Economic Co-operation and Development (OECD). "Principles on Artificial Intelligence." OECD, May 2019. Adopted by the G20.

Office of the Comptroller of the Currency. "Comptroller's Handbook: Corporate and Risk Governance." OCC, July 2019 (updated).

The six governance principles (proportionality, transparency, fairness, accountability, privacy, safety) are synthesized from NIST AI RMF, OECD AI Principles, and existing federal banking agency guidance. They represent a practitioner's integration of multiple frameworks.

Chapter 4: The Regulatory Landscape

Model Risk Management

Board of Governors of the Federal Reserve System, Office of the Comptroller of the Currency, Federal Deposit Insurance Corporation. "SR 26-2: Revised Guidance on Model Risk Management." April 17, 2026. [Supersedes SR 11-7, April 2011, and SR 21-8, April 2021.] The full text is available on the Federal Reserve's public website under Supervision and Regulation Letters.

Board of Governors of the Federal Reserve System, Federal Deposit Insurance Corporation, Financial Crimes Enforcement Network, National Credit Union Administration, Office of the Comptroller of the Currency. "Interagency Statement on Model Risk Management for Bank Systems Supporting Bank Secrecy Act/Anti-Money Laundering Compliance." April 9, 2021. [Superseded by SR 26-2, April 2026.] (Referenced in Section 4.3.)

Office of the Comptroller of the Currency. "OCC Bulletin 2011-12: Sound Practices for Model Risk Management: Supervisory Guidance on Model Risk Management." April 4, 2011. [Replaced by OCC Bulletin 2026-13, April 2026.]

The definition of "model" referenced in Section 4.1 reflects SR 26-2's revised language, which adds the qualifier "complex" to the original SR 11-7 definition and explicitly excludes simple arithmetic calculations and deterministic rule-based processes from model risk management scope.

Fair Lending and Anti-Discrimination

Equal Credit Opportunity Act (ECOA), 15 U.S.C. § 1691 et seq.

Regulation B (Equal Credit Opportunity), 12 C.F.R. Part 1002.

Fair Housing Act, 42 U.S.C. § 3601 et seq.

Consumer Financial Protection Bureau. "CFPB Circular 2022-03: Adverse Action Notification Requirements in Connection with Credit Decisions Based on Complex Algorithms." May 26, 2022. [Withdrawn May 2025. Underlying ECOA/Reg B adverse action requirements remain in effect.]

Consumer Financial Protection Bureau. "CFPB Circular 2023-03: Adverse Action Notification Requirements and the Proper Use of the CFPB's Sample Forms." September 19, 2023. [Withdrawn May 2025. Underlying ECOA/Reg B adverse action requirements remain in effect.]

BSA/AML

Bank Secrecy Act, 31 U.S.C. § 5311 et seq.

Financial Crimes Enforcement Network (FinCEN). "Alert on Fraud Schemes Involving Deepfake Media Targeting Financial Institutions." FIN-2024-Alert004, November 13, 2024.

Federal Financial Institutions Examination Council (FFIEC). "BSA/AML Examination Manual." FFIEC, 2024 (updated).

Third-Party Risk Management

Office of the Comptroller of the Currency. "OCC Bulletin 2013-29: Third-Party Relationships: Risk Management Guidance." October 30, 2013 (rescinded June 6, 2023 and replaced by the 2023 Interagency Guidance, OCC Bulletin 2023-17).

Board of Governors of the Federal Reserve System, Federal Deposit Insurance Corporation, Office of the Comptroller of the Currency. "Interagency Guidance on Third-Party Relationships: Risk Management." June 6, 2023.

Federal Deposit Insurance Corporation. "FIL-44-2008: Guidance for Managing Third-Party Risk." June 6, 2008 (superseded by 2023 interagency guidance).

Data Privacy

Gramm–Leach–Bliley Act (GLBA), 15 U.S.C. §§ 6801–6827.

Regulation P (Privacy of Consumer Financial Information), 12 C.F.R. Part 1016.

Emerging AI Regulation

Executive Order 14110, "Safe, Secure, and Trustworthy Development and Use of Artificial Intelligence." October 30, 2023. [Revoked by Executive Order 14148, January 20, 2025. Replaced by Executive Order 14179, "Removing Barriers to American Leadership in Artificial Intelligence," January 23, 2025.]

Cyber Risk Institute, Financial Services Sector Coordinating Council, and U.S. Department of the Treasury. "Financial Services AI Risk Management Framework." February 2026.

European Parliament and Council of the European Union. "Regulation (EU) 2024/1689: Laying Down Harmonised Rules on Artificial Intelligence (AI Act)." Official Journal of the European Union, July 12, 2024.

National Conference of State Legislatures. "Artificial Intelligence 2024 Legislation." NCSL, 2024 (updated).

State AI legislation is evolving rapidly. Readers should consult their state legislature's website and legal counsel for current requirements applicable to their institution.

Chapter 5: AI Ethics, Bias, and Fair Lending

Consumer Financial Protection Bureau. Final rule removing the disparate impact standard from Regulation B. Federal Register, April 22, 2026 (FR Doc. 2026-07804); effective July 21, 2026. (Referenced in Sections 4.2, 5.2, and Case Study 3.)

Consumer Financial Protection Bureau. "CFPB Circular 2022-03: Adverse Action Notification Requirements in Connection with Credit Decisions Based on Complex Algorithms." May 26, 2022. [Withdrawn May 2025. Underlying ECOA/Reg B adverse action requirements remain in effect.]

Equal Credit Opportunity Act (ECOA), 15 U.S.C. § 1691 et seq. See especially § 1691(d) regarding adverse action notices.

Regulation B, 12 C.F.R. § 1002.9 (Notifications).

National Institute of Standards and Technology (NIST). "AI 100-2e2025: Adversarial Machine Learning: A Taxonomy and Terminology of Attacks and Mitigations." NIST, March 2025.

National Institute of Standards and Technology (NIST). "AI 600-1: Artificial Intelligence Risk Management Framework: Generative Artificial Intelligence Profile." NIST, July 2024.

Chapter 6: Designing Your AI Governance Structure

The Institute of Internal Auditors (IIA). "The IIA's Three Lines Model: An Update of the Three Lines of Defense." IIA, July 2020.

Office of the Comptroller of the Currency. "Comptroller's Handbook: Corporate and Risk Governance." OCC, July 2019 (updated).

Federal Financial Institutions Examination Council (FFIEC). "IT Examination Handbook: Management." FFIEC, November 2015 (updated). The FFIEC has been incorporating AI/ML-specific examination procedures into its handbook updates; banks should review the most current version for examination expectations specific to AI governance.

Chapter 7: The AI Policy Suite

National Institute of Standards and Technology (NIST). "Artificial Intelligence Risk Management Framework (AI RMF 1.0)." NIST AI 100-1. US Department of Commerce, January 2023.

National Institute of Standards and Technology (NIST). "NIST Cybersecurity Framework (CSF) 2.0." NIST, February 2024.

Federal Financial Institutions Examination Council (FFIEC). "IT Examination Handbook: Information Security." FFIEC, September 2016 (updated).

The seven-policy suite in this chapter is the author's synthesis based on regulatory expectations across SR 26-2 (and its predecessor SR 11-7), NIST AI RMF, interagency third-party risk guidance, and FFIEC IT examination standards. No single regulatory source prescribes this exact policy structure.

Chapter 8: AI Risk Assessment: The Four Tiers

Board of Governors of the Federal Reserve System, Office of the Comptroller of the Currency, Federal Deposit Insurance Corporation. "SR 26-2: Revised Guidance on Model Risk Management." April 17, 2026. (Validation and Governance sections.)

National Institute of Standards and Technology (NIST). "AI RMF Playbook." NIST, January 2023.

Office of the Comptroller of the Currency. "Comptroller's Handbook: Model Risk Management." OCC, August 2021 (rescinded by OCC Bulletin 2026-13, April 2026).

The four-tier risk classification system (Minimal, Moderate, High, Critical) and the five-dimension scoring matrix are the author's original tools, informed by the tiered approach in NIST AI RMF and the materiality-driven principles in SR 26-2.

Chapter 9: Governing the AI You Buy

Board of Governors of the Federal Reserve System, Federal Deposit Insurance Corporation, Office of the Comptroller of the Currency. "Interagency Guidance on Third-Party Relationships: Risk Management." June 6, 2023.

Office of the Comptroller of the Currency. "OCC Bulletin 2013-29: Third-Party Relationships: Risk Management Guidance." October 30, 2013 (rescinded June 6, 2023 and replaced by the 2023 Interagency Guidance, OCC Bulletin 2023-17).

Federal Deposit Insurance Corporation. "Technology Service Provider Examination Program." FDIC, ongoing.

Board of Governors of the Federal Reserve System, Office of the Comptroller of the Currency, and Federal Deposit Insurance Corporation. "Third-Party Risk Management: A Guide for Community Banks." May 2024 (Federal Reserve SR 24-2 / CA 24-1; OCC Bulletin 2024-11).

Kikuchi, Tatsuru. "The Innovation Tax: Generative AI Adoption, Productivity Paradox, and Systemic Risk in the U.S. Banking Sector." Working paper, February 2026.

Chapter 10: Technical Controls for AI Systems

CB Financial Services, Inc. "Form 8-K, Item 1.05." U.S. Securities and Exchange Commission, May 7, 2026. Disclosure of a material incident involving non-public customer information shared with an unauthorized artificial-intelligence application; reported by American Banker and TechCrunch, May 2026.

National Institute of Standards and Technology (NIST). "NIST Cybersecurity Framework (CSF) 2.0." NIST, February 2024.

Federal Financial Institutions Examination Council (FFIEC). "IT Examination Handbook: Information Security." FFIEC, September 2016 (updated).

Federal Financial Institutions Examination Council (FFIEC). "IT Examination Handbook: Architecture, Infrastructure, and Operations." FFIEC, June 2021.

MITRE Corporation. "ATLAS (Adversarial Threat Landscape for AI Systems)." MITRE, ongoing. Available at atlas.mitre.org.

The AI-specific threat vectors described in Section 10.6 (data poisoning, adversarial attacks, prompt injection, model theft, supply chain compromise) are categorized based on the MITRE ATLAS framework and NIST AI RMF threat taxonomy.

Chapter 11: Governing Agentic AI

Draws on the model-risk guidance cited for Chapters 4 and 8 and the third-party guidance cited for Chapter 9. Primary sources: NIST AI 600-1, Artificial Intelligence Risk Management Framework: Generative Artificial Intelligence Profile (2024); OWASP, Top 10 for Large Language Model Applications (the source of the term "excessive agency").

Chapter 12: The AI Governance Committee

The Institute of Internal Auditors (IIA). "The IIA's Three Lines Model: An Update of the Three Lines of Defense." IIA, July 2020.

The model quarterly meeting agenda, dashboard design, and effective challenge frameworks in this chapter are the author's original practitioner tools, informed by IIA governance standards and the author's experience chairing IT and risk committees at a community bank.

Chapter 13: The Board's Role in AI Governance

Office of the Comptroller of the Currency. "Comptroller's Handbook: Corporate and Risk Governance." OCC, July 2019 (updated). See especially sections on board oversight responsibilities.

Federal Deposit Insurance Corporation. "Risk Management Manual of Examination Policies: Section 6.1, Management." FDIC, ongoing.

The discussion of the business judgment rule in Section 13.4 reflects general principles of US corporate law as applied to bank directors. This section does not constitute legal advice. Banks should consult qualified legal counsel regarding director liability and the business judgment rule as applied in their jurisdiction.

Chapter 14: Internal Audit and AI

The Institute of Internal Auditors (IIA). "Global Internal Audit Standards." IIA, January 2024 (effective January 2025).

The Institute of Internal Auditors (IIA). "Artificial Intelligence Auditing Framework." IIA, September 2024.

Federal Financial Institutions Examination Council (FFIEC). "IT Examination Handbook: Audit." FFIEC, April 2012 (updated).

Chapter 15: AI Use Cases Across the Bank

Board of Governors of the Federal Reserve System, Office of the Comptroller of the Currency, Federal Deposit Insurance Corporation. "SR 26-2: Revised Guidance on Model Risk Management." April 17, 2026. (Applicable to credit scoring and fraud detection models.)

Consumer Financial Protection Bureau. "Circular 2022-03." (Applicable to AI-driven adverse action notices in Section 15.1.) [Withdrawn May 2025. Underlying ECOA/Reg B adverse action requirements remain in effect.]

Financial Crimes Enforcement Network (FinCEN). "Alert on Fraud Schemes Involving Deepfake Media Targeting Financial Institutions." FIN-2024-Alert004, November 13, 2024. (Applicable to Section 15.9.)

Computer Services, Inc. (CSI). "2026 Banking Priorities." CSI, 2026. (AI-enhanced social engineering statistic in Section 15.8.)

Office of the Comptroller of the Currency, Board of Governors of the Federal Reserve System, Federal Deposit Insurance Corporation, National Credit Union Administration, Consumer Financial Protection Bureau, Federal Housing Finance Agency. "Quality Control Standards for Automated Valuation Models (Final Rule)." 2024. (Referenced in Section 15.1.)

Consumer Financial Protection Bureau. "Policy Statement on Abusive Acts or Practices." April 3, 2023. [Withdrawn May 2025; the statutory UDAAP prohibition remains in effect.] (Applicable to chatbot UDAAP considerations in Section 15.3.)

Chapter 16: Building an AI-Aware Culture

Federal Financial Institutions Examination Council (FFIEC). "IT Examination Handbook: Management." FFIEC, November 2015 (updated). See sections on training and awareness programs.

The training matrix, change management frameworks, and innovation intake process in this chapter are the author's original practitioner tools.

Chapter 17: Future-Proofing Your AI Governance

European Parliament and Council of the European Union. "Regulation (EU) 2024/1689: Laying Down Harmonised Rules on Artificial Intelligence (AI Act)." Official Journal of the European Union, July 12, 2024.

National Conference of State Legislatures. "Artificial Intelligence 2024 Legislation." NCSL, 2024 (updated).

National Institute of Standards and Technology (NIST). "AI 600-1: Artificial Intelligence Risk Management Framework: Generative Artificial Intelligence Profile." NIST, July 2024.

The four-stage governance maturity roadmap (Foundational, Evolving, Established, Advanced) is ***original to the author****. It is informed by the maturity concepts in NIST AI RMF and the FFIEC Cybersecurity Assessment Tool (sunset in August 2025), and the synthesis is the author's own.*

General References

The following sources informed the book broadly and are not tied to a specific chapter:

National Institute of Standards and Technology (NIST). "Artificial Intelligence Risk Management Framework (AI RMF 1.0)." NIST AI 100-1. US Department of Commerce, January 2023.

Board of Governors of the Federal Reserve System, Federal Deposit Insurance Corporation, Office of the Comptroller of the Currency. "Request for Information and Comment on Financial Institutions' Use of Artificial Intelligence, Including Machine Learning." 86 Fed. Reg. 16837 (March 31, 2021).

Federal Financial Institutions Examination Council (FFIEC). "IT Examination Handbook." Multiple booklets. FFIEC, various dates.

Office of the Comptroller of the Currency. "Comptroller's Handbook." Multiple booklets. OCC, various dates.

Independent Community Bankers of America. Various publications on technology, risk management, and AI in community banking. ICBA, ongoing.

Federal Bureau of Investigation, Internet Crime Complaint Center (IC3). "2025 Internet Crime Report." FBI, 2026.

Financial Crimes Enforcement Network (FinCEN). "Alert on Fraud Schemes Involving Deepfake Media Targeting Financial Institutions." FIN-2024-Alert004, November 2024. (SAR keyword: FIN-2024-DEEPFAKEFRAUD.)

Financial Crimes Enforcement Network (FinCEN). "Financial Trend Analysis: Elder Financial Exploitation." FinCEN, April 2024.

Board of Governors of the Federal Reserve System. "Synthetic Identity Fraud in the U.S. Payment System." FedPayments Improvement, 2019–2020.

Consumer Financial Protection Bureau. "Electronic Fund Transfer Act (Regulation E)." 12 CFR Part 1005.

Consumer Financial Protection Bureau v. Early Warning Services, LLC, et al. Filed December 2024; voluntarily dismissed March 2025.

Payment Systems Regulator (United Kingdom). "Authorised Push Payment (APP) Scams Reimbursement Requirement." Effective October 7, 2024.

Deloitte Center for Financial Services. "Generative AI Is Expected to Magnify the Risk of Deepfakes and Other Fraud in Banking." Deloitte, 2024.

Regulatory guidance and legislation cited in this book are current as of publication in 2026. The AI regulatory landscape is evolving rapidly. Readers should verify the current status of all cited guidance through agency websites, legal counsel, and industry associations.

About the Author

JOHN HUMPHREY JR.

John Humphrey Jr. is Vice President of Information Technology and Information Security Officer at First Resource Bank, a $1 billion community bank in Minnesota. He leads the bank's technology, cybersecurity, and risk-management agenda—setting strategy, directing operations, and advising executive leadership and the board of directors on cyber risk, regulatory compliance, and emerging technologies, including the governance of artificial intelligence. He chairs the bank's IT Steering Committee and serves on the Enterprise Risk Management Committee. This book began as a problem on his desk. As the person accountable for aligning technology strategy with business growth, regulatory expectations, and enterprise risk, he kept encountering a question no resource answered: how does a community bank—with finite staff, a constrained budget, and a full slate of existing obligations—build an AI governance program that holds up? The mega-bank playbooks didn't fit. The vendor whitepapers were thin. The regulatory guidance assumed a dedicated model risk team. What community banks needed was a guide written by someone who lives their constraints. This book is the result.

Before the VP role, John served as Information Technology Officer at First Resource Bank, where he built the bank's information security and governance programs from the ground up: control frameworks aligned to NIST, FFIEC, and GLBA; FDIC examination and external audit coordination; the enterprise vendor and third-party risk management program; and the governance foundations for emerging technologies, including responsible AI adoption. His broader portfolio spans security governance, incident response, business continuity, and enterprise-scale technology transformation.

John speaks regularly on governance and cybersecurity, most often on the discipline that defines his work: translating complex technology risk into decisions a board can own.

He lives in Wisconsin, where his faith and his family are the center of everything. The best days are simple ones: reading books to his nieces and walks with his pups. The best trip of his life was less simple—taking his mother and little sister to Italy. He trained as an operatic tenor long before he trained in risk, and the first discipline never left the second; classical music is still how he thinks a hard thing through. The pairing is less unlikely than it sounds. A score and a governance program are both exacting work, and both fail at the same place—the moment no one is truly listening.

He remains committed to a simple conviction: community banks can adopt artificial intelligence confidently, responsibly, and in a manner worthy of the trust their communities place in them.

From the Author

A NOTE ON WHAT'S NEXT

When I started writing this book, I set out to answer a single question: What does practical AI governance actually look like for a community bank—a bank like mine, where the IT leader is also the CISO, where the governance committee meets in the same conference room as the loan committee, and where the board members coach Little League with the borrowers? The trillion-dollar institutions with hundred-person model risk teams have their own answers. This book is for the rest of us.

The answer turned out to be bigger than one book.

This volume gives you the complete governance program—the principles, the structure, the policies, the risk methodology, and the playbooks for every role in the organization. It is designed to stand alone as a comprehensive reference. But as I wrote it, I kept finding areas where the guidance deserved a deeper, more specialized treatment than a single book could provide. The vendor chapter wanted to become its own book. The appendix frameworks wanted step-by-step implementation walkthroughs. The board chapter wanted to be something a director could finish in an evening and hand to a colleague the next morning.

So I'm building a series.

Volume II is already underway, and it goes deeper where this book went wide:

AI Vendor Management for Community Banks

Vendor governance is where community banks spend the vast majority of their AI governance energy—Chapter 9 could only open the subject. Volume II goes vendor-by-vendor through the core processors, loan origination platforms, digital banking solutions, and compliance tools community banks actually run—assessment guides, contract playbooks, and monitoring plans for each—organized around the questions an examiner asks about the AI you buy. If your biggest AI governance challenge is managing what your vendors are doing, the next book is for you.

And because an examination will not wait for a series to finish, a compact companion ships alongside it:

AI Examination and Audit Preparedness for Community Banks

A short book by design—built for the months between the examination letter and the exit meeting. It walks the first-day request list, the evidence binder, the vendor-AI file examiners probe hardest, the internal audit program, and what happens when a finding lands, with worked artifacts for every stage. Its premise is this book's premise: examiners examine substance, and documentation is substance made visible.

Further volumes will follow where readers tell me the need is greatest—AI cybersecurity and resilience, board-level governance, and the agentic frontier. I announce each volume on LinkedIn and through the reader list at evenstaradvisorygroup.com. If this book has been useful to you, I'd be grateful if you'd share it with a colleague—the community banker down the road who is wrestling with the same questions you were when you picked it up.

Thank you for trusting me with your time. Now go build something your community can be proud of.

John Humphrey Jr.
Hudson, Wisconsin
2026

The Community Bank AI Governance Toolkit

The complete working system behind this book—27 editable templates and three maintenance tools, organized the way your governance program actually runs.

The companion implementation package—Word policies, Excel workbooks with built-in logic (automatic risk tiering, a four-fifths AIR calculator, a stage-gate maturity scorecard), and a board-ready PowerPoint deck—is organized into seven groups that follow your governance life cycle, from the first policy you adopt to the audit evidence you hand an examiner.

GROUP	WHAT'S INSIDE
Foundation & Governance (7)	AI Governance Policy · Committee Charter + RACI · Generative AI Acceptable Use · Data Governance for AI · Vendor & Third-Party AI Risk (clause library + no-train DPA) · AI Incident Response Plan · AI Risk Appetite Statement
Discovery, Intake & Inventory (5)	Shadow-AI Discovery Survey · Use-Case Intake + Evaluation Scorecard · AI Inventory & Classification Register (automatic tiering) · Interim Acceptable-Use Memo · 90-Day Quick Start Tracker
Assessment & Validation (2)	AI Risk & Impact Assessment (with model card) · Model Validation Report
Vendor & Data (3)	Vendor AI Due Diligence Questionnaire + SOC 2 checklist · Vendor Monitoring + Concentration Register · AI Data-Flow / NPI Map + Data Readiness Scorecard
Operating Logs & Evidence (4)	Model Performance Monitoring Log · Human-Override & Exception Log · Change-Management & Re-Validation Log · Decommissioning Checklist

GROUP	WHAT'S INSIDE
Specialized Risk (3)	Fair-Lending / Bias-Testing Protocol + AIR Calculator + Test Record · Deepfake & AI-Fraud Red-Flag Checklist · Agentic-AI Guardrail Checklist
Oversight, Audit & People (3)	Board AI Reporting Pack (deck + director question card + maturity assessment) · AI Internal Audit Work Program (+ regulatory mapping, evidence binder, testing workpaper) · AI Training Rollout (+ attestation + awareness deck)

Plus Three Maintenance Tools that Keep the System Current:

Authority Register

✦

Definitions & Conventions Reference

✦

Regulatory-Change Monitoring Procedure

Every file is ready for your team to complete, brand, and deploy.

Scan the code above, or visit
evenstaradvisorygroup.com/toolkit

www.ingramcontent.com/pod-product-compliance
Lightning Source LLC
LaVergne TN
LVHW060022210926
848481LV00013B/3764
* 9 7 9 8 9 9 5 6 4 3 7 0 8 *